S0-AQI-720

WINDMILLS & WATERMILLS

WINDMILLS & WATERMILLS

JOHN REYNOLDS

PRAEGER PUBLISHERS
New York

To Joan

Published in the United States of America in 1970
by Praeger Publishers, Inc.
111 Fourth Avenue, New York, N.Y. 10003

Second printing, 1975

Library of Congress Catalog Card No. 71-101254

ISBN 0-275-49690-2
ISBN 0-275-63650-X pbk.

Designed by Norman Reynolds

Printed in the United States of America

Contents

Introduction

For thousands of years the wheel expended man's useful effort, taxing both muscles and patience in that caricature of perpetual motion the treadmill. Small wonder the reverence accorded to wind and water wheels that used as prime movers earth's natural resources, the inorganic force of wind and the kinetic energy of falling water. Such resources were replenished supernaturally as rain clouds passed overhead blown by the wind.

A gentle period of industrial revolution unfolded, marred only by social discord, the product of tensions experienced by displaced hand workers as machines became more successful. By fast flowing streams, and on exposed hillocks, strange noises attended the new prime movers—'the rush of circling sails' and the 'clucking of wooden gears'. Dramatically functional building types evolved using, for the most part, the vernacular style. Their design was invariably subservient to the machine, the economic housing of sturdy wooden works being paramount.

The ancient craft of milling was handed down from father to son, giving to the mill a perpetuity that apparently no force could end. Other changes might be seen in town and village, but the mill wheel always turning formed a constant focus of interest in an otherwise serene landscape.

As engineering techniques became more refined, efficiencies averaging 20 hp for the water wheel and 14 hp for the windmill were achieved, but the shadow of the more powerful heat engine by now loomed large. If it is true the discovery of new sources of power has been the basis for the progress of civilisation, and the rate of progress has been determined by the amount of energy available to man, then the superseding of wind and water power must surely be celebrated. But it is an inescapable fact that the harnessing of the chemical forces of fire and steam to drive engines and turbines has proved wasteful and has polluted unimaginably our atmosphere.

The disturbance of a rural pattern of industry determined largely by geomorphic considerations but with huge advantages in terms of amenity became rapidly apparent as industry hustled to the town to be coupled up to massive steam engines. Only the tide mill and hydroelectric station survive as useful wonders of a bygone revolution.

In this absorbing and most dedicated study, John Reynolds traces the development of mills and milling and explains the minutiae of milling machinery. His isometric drawings, factual records of these impressive machines, make an unique contribution explaining graphically the function of each component. The many photographs further illustrate mechanical as well as building arrangements, and show mill exteriors in their context of the countryside. Examples have been drawn from all parts of the world and each chapter gives international coverage of mill types.

The glossary of terms is an invaluable guide in analysing the component parts of these remarkable machines, and in understanding the *mystique* of milling. The engaging social content and the deep sincerity of the work as a whole combine to make this a volume to be enjoyed without detracting from its basic value as an important work of reference.

With the dwindling of coal and oil resources, the re-discovery of this vanished rural culture may serve to remind us of the present possibilities of super efficient tide mills, of the solar mill and other machines less damaging to the earth's atmosphere.

David Braithwaite, London SW11

Preface

This book attempts a brief survey of the first engines devised by man. For centuries the watermill and windmill were the only complex machines in existence, and it was not until the steam engine released man from his ancient dependence on natural sources of power that their importance began to wane. The village had possessed a mill long before the parish church could boast a clock, and the country mill was, in fact, to remain in commercial use until the close of the nineteenth century. Since then the decline has been rapid. A few mills are still at work, and others have been converted; but most stand idle, and their numbers dwindle yearly. It is sad to see buildings which have served the community for so long swept away in the course of a few careless years. The Society for the Protection of Ancient Buildings has done much to foster public interest in the cause of preservation. Certain mills are maintained by the National Trust, others by the Department of the Environment, and preservation schemes have lately been undertaken by Local Authorities in various parts of the country. But a great volume of work is also being done voluntarily, by groups of dedicated enthusiasts, determined to preserve these tangible links with the vanished world of our fathers. Their efforts deserve wholehearted support, and I hope that this book may help to focus attention on their work.

The mill was a uniquely functional building in which structure and machinery were intimately combined. It is often difficult, when clambering from floor to floor, to appreciate the overall arrangement, or even to make a satisfactory photographic record. It has, therefore, been my object to produce a set of measured drawings showing the layout of machinery in various types of mill, and to supplement these with photographs illustrating points of detail.

To demonstrate the very considerable size attained by many of these engines, a drawn scale, divided in metres, has been incorporated in each illustration. The drawings are in isometric projection, and this scale also applies, therefore, to horizontal dimensions, which run at an angle of 30° with the top and bottom of the page. In general no attempt has been made at restoration. It is felt that a literal record of a surviving machine is perhaps more useful than an impression of a 'typical' example. Mills and their machinery were constantly subject to renewal and repair, and this is reflected in the drawings, many of the installations showing evidence of late additions. For the sake of clarity it has been necessary to adopt some fairly ruthless conventions. In illustrating machinery, structural members have in general been omitted, and the cut-away technique has been avoided as far as possible. Where missing parts have been restored, the fact is noted in the caption.

First and last, the mill provided man with the flour for his daily bread; but in the course of centuries it performed many other functions, and these are briefly reviewed in the following pages. In general, buildings which were never commonly known as mills have been excluded, and this accounts for the omission of those splendid and spectacular descendants of the ancient mill, the hydro-electric power stations.

Similar problems of selection arise in dealing with the machinery of the corn mill. With the passage of time industrial processes tended to become ever more complex, requiring secondary machinery unknown to the medieval miller. Roller mills were installed in place of the traditional stones, and elaborate devices for the sifting, grading and purifying of flour came into general use. Textile mills were similarly transformed. The production of cloth involved a whole series of related trades, the more complicated of which were only adapted to water power at a comparatively late date. In the space available it would be hopeless to essay a detailed description of this formidable array of specialised machinery, much of which could in any case be considered more typical of the steam powered factory than the watermill. The aim has therefore been to single out those ancient and time honoured devices, like the stones themselves, which performed the basic function of the mill.

I must thank the many people who have made this book possible, all of whose names I cannot hope to record. I am particularly indebted to Mr. E. Moss of Whitchurch, Mr. S. Norris of Beaulieu, Mr. M. Potter of Ecchinswell, and Mr. R. Wootton of Herne, who received me with unfailing courtesy and kindness, allowing me to work for days together in their mills and to reproduce the resulting drawings. Of the other mills studied in detail some were standing in retirement, some had been restored, and others were in course of restoration. In every case I received the same kind treatment, and my thanks are due to Mr. A. Hill of Hockley, Mr. B. Jones of Blaenau Ffestiniog, Mr. A. Percival and the

Faversham Society, Mrs. M. Pyne, Mr. R. Barron and the Finch Foundry Trust of Sticklepath, Mr. J. Wright and the Cambridge Preservation Society, the East Suffolk County Council, the Kent County Council and the Welsh Folk Museum, St. Fagans, Cardiff.

Other kind people, and organisations, have made photographs available, or have allowed me access for photography, and their help is acknowledged in the list of illustrations. There remain those who have given me help and advice of a more general nature, of whom I would mention Mr. D. Bignell and Mr. C. Burrell, Mr. F. Cottrill of the Winchester City Museums, Dr. E. Course of Southampton University, Miss F. Veneman and Miss E. Sprey of the Dutch Windmill Society, Mr. E. Evans, Mr. D. Thurlow, Mr. D. Trussler and Mr. J. Turner. I am especially grateful to Mr. Hakon Eles (and the Nordiska Museet, Stockholm) and to Mr. Stewart Cruden (and the Department of the Environment) for the drawings of horizontal mills which illustrate chapter 4.

Anyone embarking on an enterprise of this sort must acknowledge his debt to previous writers on the same subject. Two authors in particular have fired the enthusiasm of a whole generation for the study and conservation of mills. They are Mr. Rex Wailes and Mr. Stanley Freese; their authoritative works are well known, and it would be difficult to over-emphasise their influence. A brief bibliography is appended, listing these and other books which deal more fully with the various aspects of milling described.

Acknowledgements

All measured drawings are by the author. Photographs have been provided by, and are reproduced by kind permission of, the following:

Mr. D. Bignell, p 67 (bottom right); Mr. D. Braithwaite, pp 97, 101 (top); the Trustees of the British Museum, pp 28 (bottom), 70 (bottom); Mr. E. Cloutman, p 24 (bottom); Mr. F. Cottrill, p 95 (right); the Public Information Office, Republic of Cyprus, pp 62 (right), 63, 156; the French Government Tourist Office, p 81; the Hungarian Tourist Information Service, pp 54, 55, 80 (bottom right), 82 (bottom); the Mansell Collection, pp 45 (bottom), 70 (top right), 94 (bottom left and right), 96, 156, 160 (left); Mr. John Marriner, p 20 (top); the Department of the Environment, pp 59, 62 (left); the Nordiska Museet, Stockholm, p 56; the Science Museum, London, pp 10, 19, 36, 40, 42 (bottom), 52, 64, 65 (top), 70 (top left), 92 (right), 105 (bottom left), 128, 130 (left), 131, 132, 133 (top left), 134, 135 (right), 150 (bottom), 162 (top and bottom left), 164 (top), 166, 167 (top left), 172 (pp 10, 40, 166 Crown Copyright); the Smithsonian Institution, pp 51, 53 (right), 65 (bottom), 66 (top and bottom left), 84 (bottom right), 95 (left); the Spanish National Tourist Office, p 80 (top right); the Ulster Museum, Belfast, p 61; the Walker Trust, St Andrew's University, p 12.

The coloured photograph of a Cretan wind pump, facing p 84, is from the Mansell Collection. That of a Minorcan tower mill, facing p 85, is by Mr. T. Daniels.

Other photographs by the author are included by kind permission of the following:

The Armfield Engineering Co., pp 66 (right), 67 (top); Mr. A. B. Dull Tot Backenhagen, pp 178 (left), 179; the Bedfordshire County Council, pp 74 (right), 103 (top right); the Dean and Chapter, Bristol Cathedral, p 72 (top left); the Cambridge Preservation Society, pp 103 (top left), 110 (bottom); Mrs D. Cullen, pp 77 (right), 100 (bottom right); the Dutch Windmill Society, pp 88, 99 (bottom right), 105 (top and bottom right), 109 (bottom left), 112 (bottom), 141, 145 (bottom), 148 (bottom left), 149 (top), 171, 172 (bottom), 176 (top right); the East Suffolk County Council, p 154 (right); the Faversham Society, p 169; the Finch Foundry Trust, pp 163, 164 (bottom), 165; Mr. F. Van Haan, pp 93 (right), 98 (top), 112 (top); Mr. B. Jones, p 120 (below right); the Kent County Council, pp 104, 110 (top); the Treasurer and Masters of the Bench of Lincoln's Inn, p 17 (top); the Revd. B. St. C. A. Maltin, p 72 (bottom right); Mr. E. Moss, pp 123 (top left), 124 (right), 126; Norwich City Museums, p 181; Mr. G. Smith, p 107 (left and top right); Mr. G. Thomas, pp 76 (left), 78 (left), 92 (left), 105 (top right); the Welsh Folk Museum, St. Fagans, Cardiff, p 119; Mr. R. Wooton, p 109 (top left and right).

The following photographs are of buildings in the care of the National Trust. Contributions towards the cost of their upkeep would be gratefully received by the Secretary to the Trust, 42 Queen Anne's Gate, London W.1:

pp 53 (left), 79 (top right), 82 (top), 86 (bottom), 93 (left), 100 (left), 103 (bottom), 107 (bottom right), 111, 114, 151 (top), 153 (right and bottom left).

1 The Origins and Development of Wind and Water Power

The age in which we live is one of constant change, as the frontiers of knowledge are continually advanced by developments in science and technology. Buildings and machines become obsolete almost as soon as they are completed, and modern man is prepared to accept, and even welcome, this state of affairs as the inevitable price of material progress. No corner of the world is so remote as to have escaped the rising wind of change, and it would seem that the process must continue, and indeed accelerate, as expanding populations force mankind to exploit the world's resources more thoroughly than ever before.

Technological development has been so rapid over the last century that although some ancient crafts survive side by side with the new technology, the watermill and windmill, man's earliest engines, have been discarded and now stand idle. However, even in the Western World there are men still alive who learned and practised the same trade as Chaucer's miller. But they are the last of the line, and when they are gone a significant link with the past will be broken.

The mill occupied its own unique place in the life of the countryside. In Britain, as the Domesday Book bears witness, its site was often as old as the parish church itself. Though its fabric might have been renewed, the ancient mill, with its ponderous machinery, remained a source of wonder to generations of children. It symbolised the continuity of life in the village, and even in the closing decades of the nineteenth century Stevenson could predict that wanderers, returning in future years to their childhood home, would still 'find the old mill wheel in motion'. In fact when those nostalgic words were written the days of the country mill were already numbered. The twentieth century has witnessed the final decline of both the watermill and the windmill, and one cannot but reflect with sadness on the extinction of an ancient craft, passed down without interruption through so many generations.

The first record of a watermill dates from the century preceding the birth of Christ. Previously man relied on his own physical strength, or on the strength of domestic animals, to perform the slow repetitive labour of grinding. The exceptional food value of the cereal grains was recognised by the earliest farmers, and the development of agriculture became inseparably linked with the growth of civilisation. Thus the wild grasses from which the modern wheats are derived were

A rotary quern, with handle and rynd restored

first cultivated in the Middle East, and excavations in Northern Iran have revealed that this process had begun at least five thousand years before the Christian era. The strains were slowly improved through centuries of selection and experiment, and wheat became established as a food crop in lands far removed from its original home. It is interesting to note that this process has so transformed the species that modern varieties are now dependent on man for their survival.

The first men to sample the seeds of wild grasses cracked them between their teeth; but it must soon have become apparent that if grains were to be eaten in quantity, some preliminary crushing process was necessary. The wheat grain or berry, in common with other cereals, consists of a starchy endosperm or kernel enclosed in a very hard outer sheath of bran. Within the sheath, close to the stalk, lies the minute embryo or germ of the future plant. Wheat is exceptionally rich in carbohydrates and also contains minerals, protein, vitamins and fat. Some of these elements occur in the bran, but this part of the grain is not readily digested by human beings.

The first farmers prepared their grain by pounding it between stones, but in the course of time this primitive crushing process was superseded by the use of the saddle quern. Here a true grinding action was produced by working a rubbing stone backwards and forwards across a saddle shaped stone base. Soon every household possessed a quern, and grinding for the family was a time consuming daily ritual performed by the womenfolk. The rich employed servants or slaves, and numerous statuettes from Egyptian tombs depict women kneeling at saddle querns, grinding meal for the use of the dead in the after-life. Many centuries later, this device in turn gave place to the rotary quern, which employed a circular stone mounted on a central pivot, swung to and fro through an arc of a circle by means of an eccentric peg or handle. The action of the crank was unfamiliar to early man and these querns did not, at first, make use of continuous rotary motion. It was from these beginnings, however, that the revolving millstone was developed, and with it the skilled craft of the miller.

Commercial milling establishments made their appearance during the Roman period and in these a larger and more sophisticated version of the rotary quern, operated by asses or slaves, was frequently employed. The revolving stone was shaped like a hollow hour glass and turned on a conical stone base. Examples of such mills survive in situ among the ruins of Pompeii, and similar remains have been excavated at many other Roman sites. It is at this point in time that the first known reference to a watermill occurs in the form of an epigram by Antipater of Thessalonica, written in about 85 BC:

> *'Cease your work, ye who laboured at the mill. Sleep now, and let the birds sing to the blood red dawn. Ceres has commanded the water nymphs to perform your task; and these, obedient to her call, throw themselves on the wheel, force round the axle tree and so the heavy mill.'*

The machine described by the poet was evidently of the type later to become known as the Greek mill. The upper stone was driven from below by a vertical spindle, on the lower end of which was mounted a wooden rotor or impeller. There was no gearing, both rotor and runner stone revolving at the same speed. Such mills may have existed before the time of Antipater, but earlier evidence is lacking and it is reasonable to assume that they first became known to the Western World at about this date. It is significant that within a few years a similar device appeared in China, where the use of a horizontal water wheel to operate bellows is recorded in AD 31. The investigation of silt deposits associated with artificial water courses in Jutland has led archaeologists to the conclusion that primitive mills existed here also, at about the time of Christ. This evidence suggests that the horizontal mill may well have been an invention of the barbarian world and that the principle was diffused from some unknown source outside the bounds of the Roman Empire.

Strabo records the existence of a watermill at the palace of Mithridates, King of Pontus, where it aroused the wonder of Pompey's conquering troops in 63 BC. It appears, however, that soon after their introduction these simple mills had been widely adopted and that when Pliny, some 140 years later, mentions the existence of mills throughout rural Italy, he is referring to the primitive Greek machine.

The second basic mill type also appeared during the Roman period, although again its ultimate origin is uncertain. This was the vertical or Roman mill, the direct forerunner of the traditional watermill which was to remain in common use throughout the West until the present century. Vitruvius, whose great work *De Architectura* was probably written some 15 years before the birth of Christ, describes the Roman mill in detail. His specification includes all the essential components of the later medieval mill—a vertical water wheel fitted with blades or paddles, a toothed pinion keyed to the same horizontal axle and a larger toothed gear wheel mounted on the vertical spindle of the millstone. Mechanically, the introduction of gearing marked a great advance over the Greek mill, allowing power to be transmitted from one plane to another, from a vertical water wheel to a horizontal runner stone. This innovation was also to permit the future use of differential speeds, although Vitruvius' description makes it clear that in the early Roman mill the runner stone, strangely, revolved more slowly than the wheel. From this it may be inferred that the Vitruvian mill was intended for use on fast falling streams.

It is a curious fact that Rome was slow to exploit the watermill, and indeed the new machine met with strong official opposition. Huge numbers of slaves were employed daily in the city's mills and it was claimed that the State would be endangered if this great labour force were to be thrown out of work. It is probable, however, that there was another deeper reason for opposition. A certain lack of interest is noticeable in the classical attitude towards machinery generally and it is at first sight remarkable that more progress was not made at this period by Greek and Roman engineers. They were well acquainted with the basic principles of gearing and the latent potential can be gauged from such devices as the astronomical computer, recovered in recent years from a wreck in the Aegean. This complex train of gears, which has been dated to the first century BC, affords compelling evidence of the technical skill available. Various devices made use of hydraulic power, but often for purposes which seem strangely frivolous to the modern mind, such as the operation of temple doors or theatrical machinery. In fact the ancients viewed the forces of nature with a deeply ingrained religious awe which tended to inhibit experiments with water power. When every river or waterfall was inhabited by its own *genius loci*, to set up a water wheel was tantamount to harnessing the gods for the menial service of man. But purely practical considerations, allied perhaps to a growing cynicism, ensured the gradual acceptance of the watermill.

Opposition to the mill ceased with the adoption of Christianity as the official religion of Rome under the Emperor Constantine, and from the fourth century onwards a number of edicts were issued to regulate the conduct of milling within the city. Millers were already long established on the Janiculum hill, and the old aqueduct of Trajan, which carried water from Lake Sabbatina, was devoted to their use. The rapid spread of the watermill at this period was not limited to the Roman Empire; according to tradition, the reign of Constantine also saw its introduction to India from Asia Minor. The earliest known Western representation of a watermill occurs in a mosaic dating from the early fifth century, found in the ruins of the Great Palace at Constantinople. Although difficult to interpret in detail, the building depicted, with its great undershot wheel, is instantly recognisable as a mill.

The earliest vertical water wheels, like that in the Byzantine mosaic, were undoubtedly of the undershot type. They were turned by the current acting against flat projecting paddles, and the velocity of the stream dictated the speed of rotation. The more efficient overshot wheel was, however, also developed during the Roman period. Where the levels of the ground allowed, water was delivered over the top of the wheel from an elevated mill race or flume. The paddles retained the falling water, making use of its weight to drive the mill.

Archaeologists have investigated the remains of various Roman mills, including an early example at Venathro near Cassino. Here the building had been engulfed by a volcanic eruption, but a matrix of lava had preserved the imprint of the undershot timber wheel. This measured slightly more than six feet in diameter, and was fitted with radial floats enclosed between deep timber rims. Another mill excavated during recent years stood in the Agora at Athens. Built during the latter part of the fifth century AD, it had been equipped with an overshot wheel driving two pairs of stones. But the most remarkable evidence has come

Mosaic at the Great Palace, Byzantium, depicting a Roman watermill, early fifth century—the earliest surviving representation of a watermill

to light at Barbegal in France, where a large multiple milling installation was built in the time of Constantine. Set on a hillside, it was served by a double mill race carrying water down the steep incline from an aqueduct. Power was derived from no less than eight pairs of overshot wheels, each driven by the tail water from the wheel above, and coupled to a pair of millstones measuring about three feet in diameter. The axles were of iron, and carried wooden gear wheels secured with run lead joints. Although no such dramatic discoveries have been made in Britain, there can be little doubt that the Romans also introduced the watermill into this distant province. The sites of two mills have been tentatively identified near Hadrian's Wall and a stone hub, recovered from the River Witham at Lincoln, probably belonged to a water wheel of the Roman period.

The basic principle of the grist mill, thus established by the Romans, was to survive for nearly two thousand years. The invention of the European windmill in the late twelfth century introduced ingenious alterations in power transmission, but the grinding process remained unaltered. The stones of the medieval mill were remarkably similar to those used in ancient Rome. The grinding surfaces were dressed with a pattern of radial grooves or 'furrows' designed to shear rather than crush the grain, which was reduced to flour in a single passage from the centre to the circumference of the stones. This process was known as 'low milling', so called because the runner stone was set low, reducing to a minimum the clearance between grinding surfaces. A skilled miller was able to obtain a high proportion of good quality flour by this method. Brown with bran and fragrant with the crushed wheat germ, it contained natural vitamins excluded from the 'pure white' flour of modern times. This was the process for which the traditional mill was designed. It was practised throughout the Middle Ages, and was to survive in the country mill until the end of the nineteenth century.

The vertical water wheel itself was not a Roman invention. There can be little doubt that it evolved directly from the noria or 'Persian wheel', employed in the great river valleys of the Near East during the classical period, and remaining in use to the present day. An Egyptian papyrus of the second century BC indicates that it was already known at that date, and must therefore claim a place among the earliest engines devised by man.

In principle the noria consisted of a wheel or chain of bailers, revolving on a horizontal axle, and set up on the bank of a stream to raise water from a lower to a higher level. It existed in several differing forms, three being distinguished by Vitruvius in his characteristically detailed description. The first of these was an endless chain of containers, passing over a driving sprocket. The second was a rigid wheel, with containers around the rim, which dipped below the surface of the water at each revolution. The third type consisted of a 'tympanum' or hollow scoop wheel, with internal divisions forming a series of water-tight compartments. The momentous innovation of automatic action was achieved in this latter version by the use of paddles around the rim of the wheel, which caused it to turn in the current. It must have been observed that wheels of the simple type showed a tendency to revolve in swiftly flowing streams, and perhaps the action of driftwood caught in the spokes suggested to some ingenious mind the idea of attaching floats. As his wheel began to turn, the inventor can have had little inkling of the importance of the train of events thus set in motion.

Where there was no flow of water available, the heavy monotonous work of turning the wheel was performed either by treadmill action or by the use of draught animals. Scoop wheels worked by slaves were used for mine drainage, and for raising water in baths and other public buildings. At Rio Tinto in Spain, the exploration of ancient copper workings revealed the remarkably well-preserved remains of an elaborate drainage system. Narrow timber wheels, arranged in pairs and linked by treadmills, were employed to lift water in stages through a vertical height of about 100 feet.

When animals were employed, it was obviously preferable for them to work in a horizontal plane, and this necessitated the introduction of gearing similar to that of the Vitruvian mill. The invention of such gearing was traditionally ascribed to Archimedes (287-212 BC), although it may have occurred even earlier. The date of its adaptation to the irrigation wheel is not known, but it is evident that the development of this device was intimately bound up with that of the watermill.

Egypt has always been uniquely dependent on irrigation, and it is not surprising that much pioneering work in the field of hydraulics was carried out in the Greek school at Alexandria. The principles enunciated by Archimedes were here developed during the second century BC by such men as Ctesibius, credited with the invention of various hydraulic devices including the force pump. The Archimedean screw or cochlea was widely employed, and Strabo records the existence of examples in the Nile delta at about the time of Christ. Whilst neither the pump nor the screw was operated mechanically during classical times, both were to be used extensively in later drainage mills.

Another machine which was to have a profound effect on the techniques of milling made its appearance during the classical period. This was the edge runner mill, commonly employed by the Romans for the production of olive oil. It consisted of a pair of broad stone wheels freely mounted on a short horizontal axle which projected on either side of a vertical shaft. Rotation of the shaft caused the wheels to revolve on a circular stone base, crushing the olives laid in their path. The mill was driven by animals or slaves,and there is no evidence to suggest the early use of water power. With its continuous rotary motion, however, the edge runner mill was admirably suited to mechanical operation, and it was very widely used for a variety of purposes in Europe from the seventeenth century onwards. It is a curious fact that in this instance again, a parallel development was taking place in China, where the use of the edge runner mill is recorded in the second century AD.

Romano-British millstones from Richborough, Kent. The larger stone is of Kentish rag, and dates from the first century

Although the grinding action of the millstone was generally adopted for the reduction of cereal grains, the simple pounding action of the pestle and mortar was well suited for some processes, particularly that of husking. It was for this purpose that the water driven tilt hammer was first developed in China, where it was employed from an early date for the hulling of rice, thus providing another basic tool for the millwright of later years.

Although water powered bellows are recorded in the first century AD in China, the surviving evidence suggests that during the classical period the water wheel was used in Europe for two purposes only; for driving corn mills,and raising water. A single puzzling reference to saw mills on the banks of the River Moselle occurs in the works of Ausonius, written during the fourth century AD, but this can possibly be discounted as a later interpolation. The latin word *mola* was applied first to the quern, the 'mill' of the Old Testament, and later to any mechanical grinding device. As water power was harnessed to perform other work, however, the precise meaning of the word was further modified to include buildings and machinery employed for a variety of industrial purposes. The word 'mill' therefore defies exact definition, and any study of windmills and watermills must inevitably include a variety of building types, designed for widely differing purposes. With the introduction of the steam engine as a source of power, the link became even more tenuous, many of the new factories retaining the ancient name. Even today this bears very different connotations in the country village and the industrial town; and the idyllic rural scenes painted by Constable, himself the son of a miller, contrast strangely with the 'dark satanic mills' of William Blake.

2 Water Wheels and Millstones:

The Application of the Vertical Water Wheel to Corn Milling

The principle of the watermill survived the Dark Ages following the fall of Rome, but evidence relating to this period is scarce. In Britain, references begin to appear in Saxon charters during the tenth century, and it is known that several mills were at work in Winchester in 983, during the reign of King Edgar. The Domesday Book of William the Conqueror, that unique survey of the resources of Britain in the years following the Conquest, records the existence of no less than 5624 mills. Some of these may well have been worked by horses or oxen, but there is no doubt that most were true watermills. This conclusion is justified by a number of factors, notably the location of the communities served, and the rents quoted for the mills themselves, which frequently included payments of eels. The eel fishery, in fact, was to supply a significant part of the miller's income for centuries to come.

During the Middle Ages, the mill belonged to the Lord of the Manor, who possessed by ancient custom 'soke rights'. All corn grown on the Manor was required to be ground at the Lord's mill, the miller deducting a proportion of flour or meal as payment for his services. This toll varied widely, but most commonly amounted to about one sixteenth of the final product of the mill. It was measured in a 'toll dish', or 'multure bowl', struck off level with a 'strike' or 'strickle'. Tenants were strictly bound to the manorial mill; they also owed service to the Lord and this could take the form of work on the repair of the structure. But feudal privileges entailed certain obligations, and the mill was expected to provide a reasonable service. In deciding on its site, the Lord was required to consider the convenience of his tenants, who were entitled to go elsewhere if the building was not properly maintained. Grain delivered to the mill was ground in strict sequence, the Lord alone having a right of precedence and paying no toll. A time limit was set for grinding; sometimes this amounted to 'a day and a night', sometimes longer, and if grain was not dealt with in the time allowed, the owner was free to take it to another mill. The activities of the miller were strictly circumscribed by law, and until the reign of Queen Elizabeth I, he was forbidden to deal in grain or meal. He was allowed to sell only the corn taken as toll, though he often supplemented his income by rearing pigs on the bran from the mill. In cities the payment of money in lieu of toll had begun as early as the year 1281, but it was

Hambleden Mill, on the Thames

not to become compulsory in Britain until the end of the eighteenth century, and even at that late date toll might still be taken with the consent of both parties.

Only ancient manorial mills were entitled to claim soke rights, and with them the valuable privilege of freedom from tithes. The creation of manors ceased early in the fourteenth century, during the reign of Edward I, and in cases of rebuilding owners would go to great lengths to prove the early foundation of their mills. Rights might be allowed to lapse, and were sometimes deliberately relinquished, but since they were never defined by statute law, they were never formally abolished. They were, nevertheless, upheld in the courts, and it was not until the nineteenth century that the system finally fell into disuse. In a number of cases towns were forced, as a last resort, to purchase their freedom. Wakefield did this in 1854, but it was not until 1871 that Bradford obtained release from a system which had long since ceased to fulfil its original purpose.

The medieval system of milling and toll was inflexible and open to abuse. It was unpopular with the peasants, who would have preferred to grind their small quantities of grain on hand mills in their own cottages. This practice did in fact continue, and on one famous occasion the Abbot of St. Albans paved a courtyard with quern stones confiscated from his mutinous tenants. This action was not forgotten,and during the Peasants' Revolt the courtyard was destroyed as a symbol of the Abbot's despotic rule.

An efficient mill was recognised as an important financial asset. Its construction and maintenance, however, involved considerable expenditure and the Lord of the Manor no doubt felt fully justified in enforcing his rights. But the system took no account of the effects of inflation and rising prices which followed years of plague or famine, and at such times many millers must have struggled to earn a living. It is small wonder that the miller acquired a reputation for dishonesty. He alone knew the quantity of meal produced from any sample of grain, and he was continually suspected of helping himself to more than his share. In the community, he occupied a curious and solitary place, the mill often standing, of necessity, away from the main settlement and the miller having no part in the daily labour in the fields. In the *Canterbury Tales*, Chaucer depicts the medieval miller as a brawny, loudmouthed rogue, unscrupulous and quarrelsome. No doubt there were such men, but equally certainly the average miller was no more and no less dishonest than his contemporaries. Strangely the unpopularity of the medieval miller does not seem to have extended to his family. Every child knew that Much the miller's son had been a trusted member of Robin Hood's valiant band; while the hero of Puss in Boots was again the son of a miller and Puss himself a mill cat.

The craft of milling was normally handed down from father to son, and certain characteristics have distinguished the miller down to the present day. He was the engineer of the community, trained from his youth to manhandle heavy loads in confined spaces. His calling required both physical strength and ingenuity, and tended to make him mentally more alert and independent than many of his rustic neighbours. Through years of experience he acquired the curiously flattened 'miller's thumb', which resulted from the constant testing of meal between the thumb and fingers of the left hand. He prided himself on his ability to gauge, by touch, the fineness of meal and the quality of grain, and his livelihood depended, to no small extent, on what Chaucer referred to as the 'thumb of gold'. According to popular lore, the honest miller could display another distinguishing mark, a tuft of hair in the palm of his hand. To this jibe the miller could only respond by pointing out that it

LEFT
Stockleigh Mill, Devon

RIGHT
Arms of Lincoln's Inn, 'Azure fifteen millrinds or, on a chief or a lion rampant purpure'

BELOW RIGHT
The tide mill on Ashlett Creek, Hampshire

required an honest man to see it. A more sinister characteristic was the 'miller's cough', a respiratory complaint caused by the inhalation of dust. This was to become sadly prevalent during the nineteenth century, until countered by the installation of efficient dust collecting machinery.

The mill of the early Middle Ages was a simple structure, often no more than a rectangular hut of timber or wattle and daub. The water wheel turned in the open air, its axle passing through the wall to drive a single pair of stones within the building. The mechanism was directly derived from that of the Roman mill described by Vitruvius. Illustrations appear in many medieval manuscripts, and indeed the type was to survive in remote areas down to the nineteenth century, the great water wheel being the characteristic feature of the building. At Durngate mill, Winchester, the name of the first miller to be recorded, in 1213, was Edwerd Alwele, Edward 'of the wheel'. By comparison with other craftsmen, millers were slow to achieve corporate status, but when separate guilds were formed, as in central Europe during the fifteenth century, the water wheel was commonly displayed upon their seals. Other components of the mill were also adopted as heraldic charges, the strong symmetrical shapes of both the millstone itself and the rynd (or *fer-de-moline*), making them ideally suited for this use. The mill rynd was the iron bridging piece which spanned the eye of the runner stone. The heraldic rynd, which resembles an italic letter 'x', represents an ancient pattern long since superseded in England but still to be seen in some continental mills. Both rynds and millstones were most commonly employed as canting symbols to represent particular surnames such as 'Mill' or 'Mylner'.

In the Middle Ages roads were few and poor, and the extent to which rivers were used for transport tends to be forgotten. Many cathedrals and churches far inland were built with stone brought up laboriously by barge along rivers which today seem too swift or too shallow for navigation. These waters came under the jurisdiction of the Lord High Admiral, who thus found himself called upon to order the destruction of mills and weirs where they threatened to impede an old established trade route. Such instructions were not always obeyed. The great weir at Chester which sweeps to this day across the River

Eling tide mill and causeway

Sluice gates at Beaulieu

A floating mill. From Machinae Novae *by F. Veranzio, c. 1590*

Dee, and for centuries provided power for the mills of that city, survives in spite of a Royal command, twice repeated, for its removal.

Mill sites once established were jealously guarded. The raising of weirs and the cutting of races inevitably affected the flow of water to other mills on the same stream, and gave rise to innumerable law suits. With the introduction of breast-shot wheels requiring a positive head of water, the number of mills which could be supported by a given stretch of river became strictly limited. Small streams were pressed into service by the building of dams and the formation of mill ponds, which allowed the miller to grind during the day, his stock of water being replenished overnight. These man-made dams had a lasting effect on the landscape, and in many places wide mill ponds and narrow flooded valleys are today the sole evidence of past industrial activity. Small seasonal streams were sometimes employed if no better source of power was available, and such sites were probably the evocatively named 'winter mills' of the Domesday survey.

Land owners were tempted to harness the vast power of the sea, and from the eleventh century onwards, tide mills were erected when suitable sites could be found. The mill recorded at Dover in the late eleventh century as causing 'disaster to vessels by the great disturbance of the sea', may well have been such a structure. Storms and high tides were a constant threat, and these tide mills seem to have been localised, in the British Isles, in areas where shelter could be found from the direct onslaught of winter gales. A typical grouping survives on the south coast, where the Isle of Wight afforded protection from the prevailing South Westerlies. Very considerable causeways were sometimes constructed to impound the water in tidal rivers and creeks, and the antiquity of these embankments is often witnessed by the fact that they carry old established roadways. At other sites a sea wall was raised to enclose a mill pond along the shore of a convenient inlet. The flood tide was allowed to run upstream unhindered, through sluice gates which were closed at high water. In some cases automatic gates were employed and these presented a curious spectacle at the turn of the tide, slowly swinging to against the ebb, and opening again as the flood tide began to make. When the water level below the mill had dropped sufficiently, the retained water was allowed to escape through the mill race. In many havens the violent flush of water from the mill served a dual purpose, helping to scour the channel down to the sea.

Tide mills were commonly powered by undershot wheels set below the floor in order to accommodate the wide tidal range of the British coast. At high water springs the tide would be lapping at the floorboards; at low water the whole of the wheel might be exposed above the muddy bed of the race. The miller's day was rigidly regulated by the tides, but he was assured of power for two periods, each of two to three hours duration. His working days were thus irregular but predictable, and this was reckoned no hardship in a coastal community, where men's lives were geared to the phases of the moon and not to the sun.

The establishment of mills on major rivers, particularly where marked seasonal or tidal changes of level occurred, presented special problems, and this fact is emphasised by the absence of Domesday mills along the tidal reaches of the Thames and Severn. In some cases mills were perched on a cluster of timber piles, and stood high and dry when the level of the river was low. Because wide variations in water level meant that a fixed wheel could seldom function at full power, limited use was made of adjustable wheels, which could be raised or lowered with the stream. Some large and powerful examples were constructed in France during the latter part of the nineteenth century, but installation was costly and they were never widely used. A simple and functional

alternative was the floating mill, and these strange ark-like vessels were well known in the cities of medieval Europe. Their invention is traditionally ascribed to the resourceful Byzantine general Belisarius, who defended Rome against the assaults of the Ostrogoths during the first half of the sixth century. When the mills were destroyed by the enemy, he ordered millwrights to install machinery in barges which were moored in the Tiber, thus maintaining a supply of flour for the defenders of the city.

Floating mills were necessarily powered by simple undershot wheels, but these remained at a fixed height above the water, and could thus operate in the most efficient manner possible. Two types of mill may be distinguished. The first was equipped with two water wheels, mounted like the wheels of a paddle steamer on either side of a double-ended barge. The second employed a single wheel, which revolved between a pair of rigidly linked pontoons. The millstones were set on a timber frame or 'hursting', driven by conventional gearing, and protected from the weather by a pitched roof. In cities, floating mills were most commonly moored beneath the arches of bridges, and in Paris the Great Bridge was their traditional location. The roadway afforded a convenient means of access, while the massive piers created a sharp increase in the speed of the current, thus providing the miller with a ready made race. In this position, however, mills constituted a spectacular hazard to shipping. 'Shooting' a bridge with a loaded barge could be a difficult and dangerous operation at the best of times, and it was frequently necessary to regulate the number of mills in the interests of navigation. Where the river flowed through open country, the miller was denied the convenience of a bridge, and floating mills were moored out in the stream, grain and meal being ferried to and fro by boat. A few remain in use today serving scattered farming communities along the shores of the River Danube below the Iron Gates.

Milling was a necessary task which had to continue in time of war, and it was not uncommon for a walled town or castle to be provided with a fortified mill. If the moat was fed with running water, the outfall presented an obvious site, and since the sluice gates formed a vital part of the defences, it was naturally vital to protect them against attack. At Southampton, the town ditch was filled by the tide, and a mill was

ABOVE
God's House Tower, Southampton: the former outfall from the town moat was under the relieving arch visible on the right

TOP
A floating mill on the Danube

Abbotstone Mill, Hampshire

The structure of a small country mill. Scale in metres

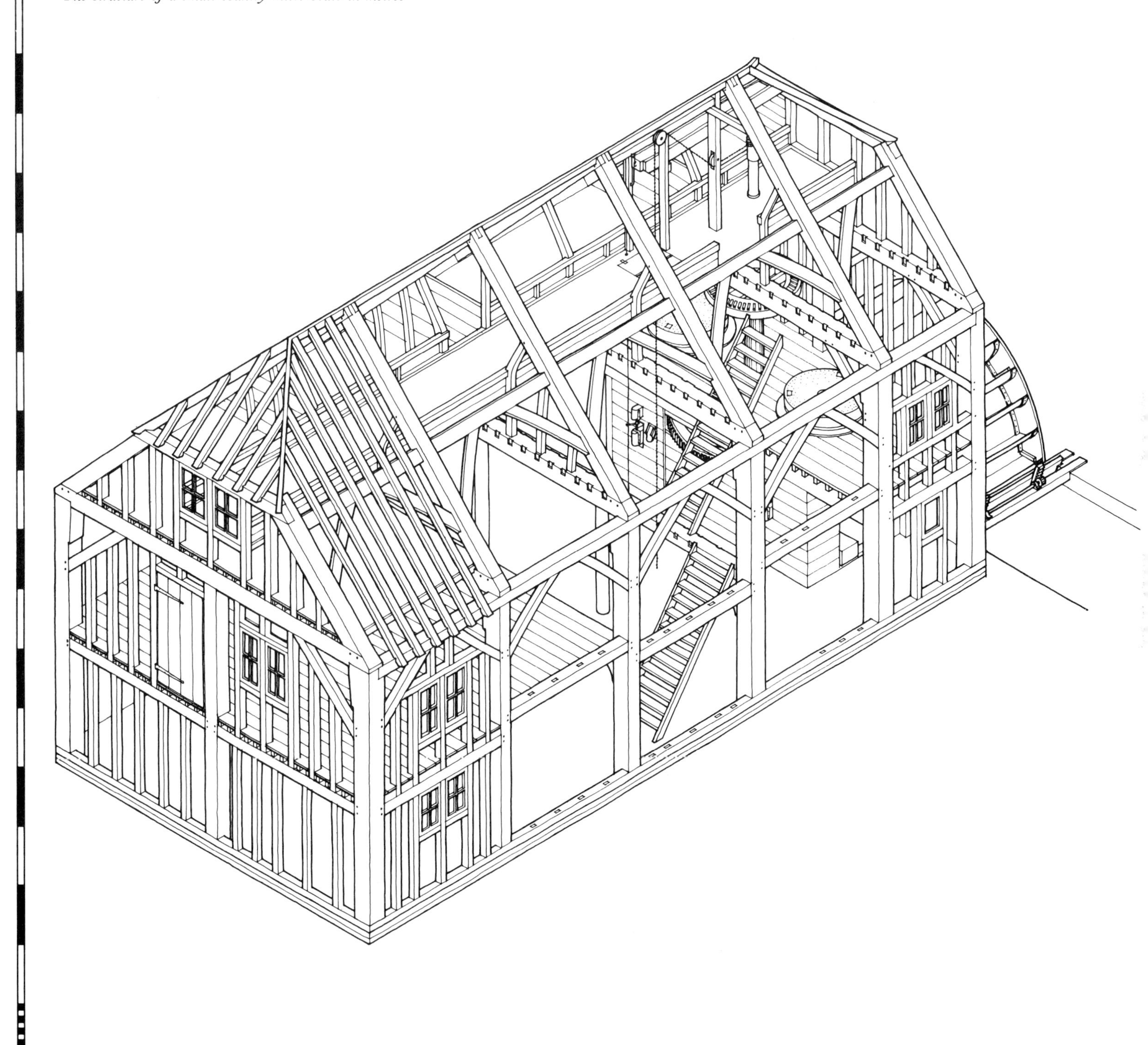

installed in God's House Tower, a massive stone barbican erected in the thirteenth century to command the moat at the seaward corner of the walls. The tower remains, and the relieving arch above the race may still be seen, but the moat itself, and the machinery of the mill, have long since disappeared.

Few structures have made more efficient use of enclosed space than the mill, and an eminently functional building type was evolved over the years. The little mill which stood at Abbotstone in Hampshire was typical of the many similar buildings which served the villages of Britain during the eighteenth and nineteenth centuries. Abbotstone mill was a simple rectangular building, its stout timber frame concealed behind a cladding of tarred weather-boarding. The roof was of Cornish slate, carefully laid in diminishing courses, and lit by dormer windows. Power was provided by an iron undershot wheel, of Poncelet type, cast in the nearby town of Alton during the late nineteenth century. This had replaced an earlier wheel, and various alterations to the building, including the provision of a wheel house, had evidently been made at the same date.

The wheel was mounted against one end of the mill, the axle passing through the external wall into the 'cog-hole', or machinery space within the ground floor. Here a frame of heavy squared timber enclosed the main gearing, and supported the stones on the floor above. The sides of this hursting were fitted with wooden shutters, secured by turnbuckles, allowing access to the gearing, and to the thrust bearing of the great vertical shaft. Standing before the hursting the miller was in full control of the grinding process. He could judge the texture of the warm meal as it fell through wooden spouts from the millstones above, and adjust the 'tentering gear' accordingly, raising or lowering the runner stones by means of hand-screws. The 'crook strings' which governed the flow of grain were also within easy reach, as was the iron hand-wheel which controlled the sluice gate, and thus the speed of the mill. Watermills were without brakes, the machinery being set in motion, or stopped, by the operation of the sluice.

The main vertical shaft rose like the mast of a ship through the height of the first floor above. It was flanked by the stones, each pair enclosed in its wooden 'tun', and surmounted by the usual horse and hopper. Also located on the stone floor were the auxiliary machines of the mill, driven by belts from pulleys mounted on a lay shaft under the ceiling. Power was supplied by the crown wheel, a stout bevel gear wheel of 'clasp-arm' construction secured with wedges to the top of the main shaft.

The principal loading doors were at first floor level, enabling sacks of grain to be lifted from wagons drawn up beside the mill. To facilitate this operation a wooden slide was hinged to the door sill, and hardwood rollers were mounted on the beams to allow the hoist chain to be led out through the doors without chafing. The whole of the roof space above was utilised for the storage of grain. Under the rafters, a series of open bins extended along each side of the building and these were served by a raised central gangway running the length of the mill. A secondary vertical shaft, mounted directly above the main shaft, and driven from it by an iron clutch, was carried up to the ridge to serve as a bollard for the sack hoist. This device was hung from a timber balance beam, and was dropped into gear by pulling on a rope passing down through the floors of the mill. The arrangement at Abbotstone was unusual, most millwrights preferring a horizontal bollard, but this was a part of the installation which tended to differ widely from mill to mill.

BELOW
Sluice gates at Terwick Mill, Sussex

Abbotstone Mill, Hampshire

The drawing shows the machinery of a traditional mill. The iron water wheel (right), by Hetherington and Parker of Alton, was installed in 1876 to replace an earlier wheel. The bevelled pit wheel (bottom centre) drives an iron wallower mounted at the foot of the main vertical shaft. Above the wallower can be seen the great spur wheel (centre) of timber 'compass arm' construction, supplying power to two pairs of millstones through wooden pinions, or stone nuts, mounted on iron spindles. These are supported on timber bridge trees (left, centre), tentering being effected by handscrews. Near the top of the main shaft is a wooden 'clasp-arm' crown wheel, and a lay shaft (top left) from which the secondary machinery of the mill is driven by belting. Above the main shaft is the vertical bollard of the sack hoist (top centre) driven by an iron clutch and suspended from a heavy balance beam. Scale in metres

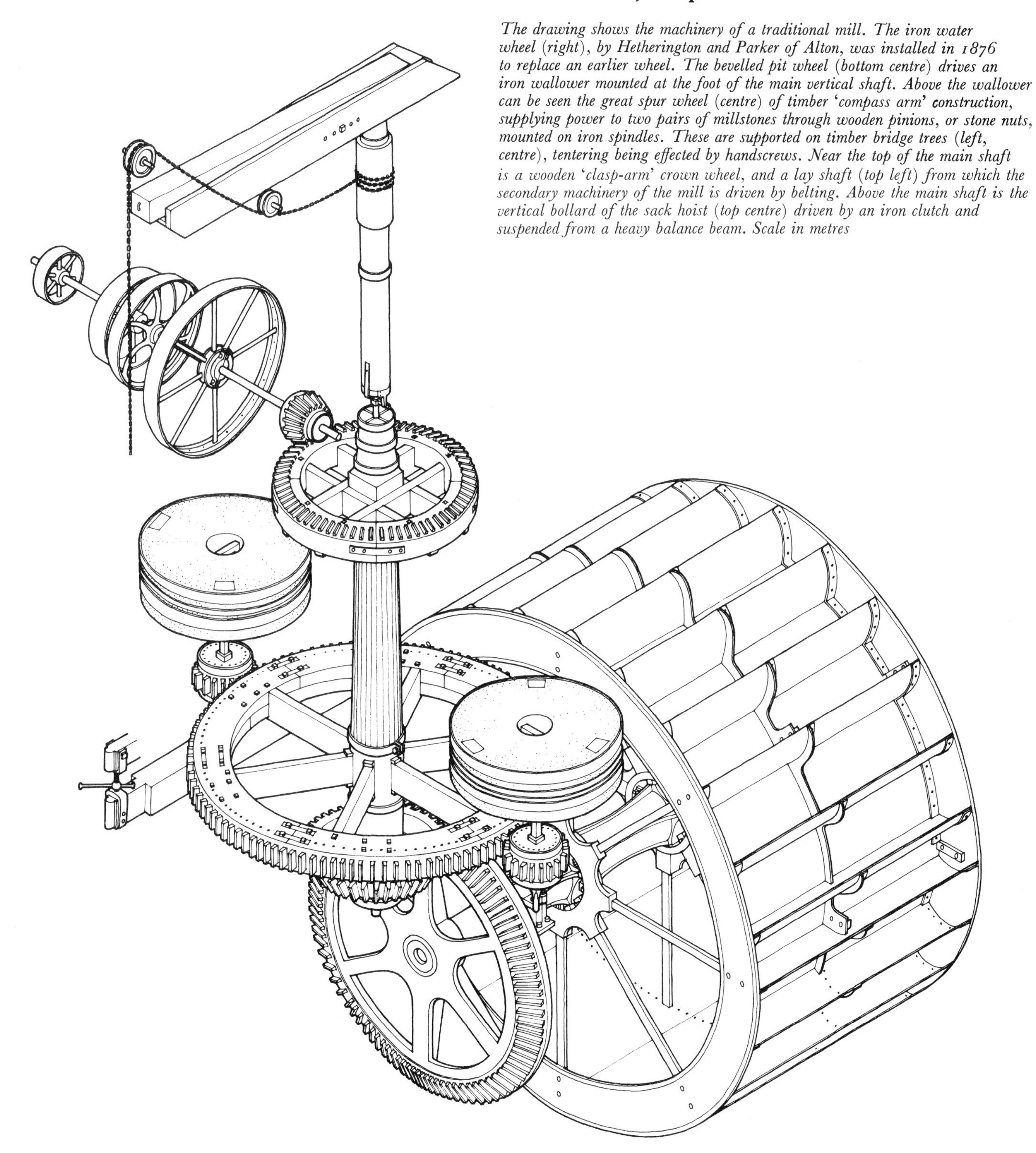

ABOVE RIGHT
The threshing mill at Hockley, Hampshire, 1803

ABOVE LEFT
Abbotstone Mill—the head of the main shaft, showing the crown wheel and pinion. The iron clutch (top centre) drove the vertical bollard of the sack hoist on the floor above

BELOW LEFT
Abbotstone Mill—the roof space, showing the raised central gangway flanked by storage bins

As the enforcement of soke rights fell into abeyance, many wealthy farmers found it profitable to operate their own mills. These might be old manorial foundations, or new buildings erected on ancient sites. During the early nineteenth century a number of farm mills were equipped to perform a new and revolutionary process, that of threshing. The threshing machine had been invented by Andrew Meikle in 1784, and an early example of a watermill employed for this purpose survives at Hockley in Hampshire. This handsome brick building bears the date 1803, and consists of a lofty barn with a projecting wing on one side housing the mill. The threshing machine itself, which occupied the central bays of the barn, has gone, but the driving machinery, powered by a later undershot wheel, survives intact. In place of the conventional vertical shaft, the pit wheel here drove a ponderous horizontal axle, which transmitted power through the length of the mill to an iron spur wheel mounted against the wall of the barn. The first floor layout resembled that of a conventional two-pair mill, the stones being driven by wooden bevel gearing from the main shaft below. A diminutive vertical shaft carried a crown wheel mounted below the tie beams, and this in turn drove the hoist on the bin floor above. This particular mill is an interesting and well documented survival. It was described at some length by Charles Vancouver, whose *Agricultural Survey of Hampshire* was published in 1813. It was then said to be capable of threshing 30 bushels of wheat in an hour, and was something of a local wonder.

Hockley Mill, Hampshire, 1803

The machinery of a country mill designed for the dual function of threshing and grinding. The threshing machine itself, which would have occupied the left foreground of the drawing, has not survived. The iron water wheel, 12 ft. 6 in. in diameter, bears the date 1888, and is a late replacement. In the drawing the short layshaft on the left of the crown wheel has been restored. Scale in metres

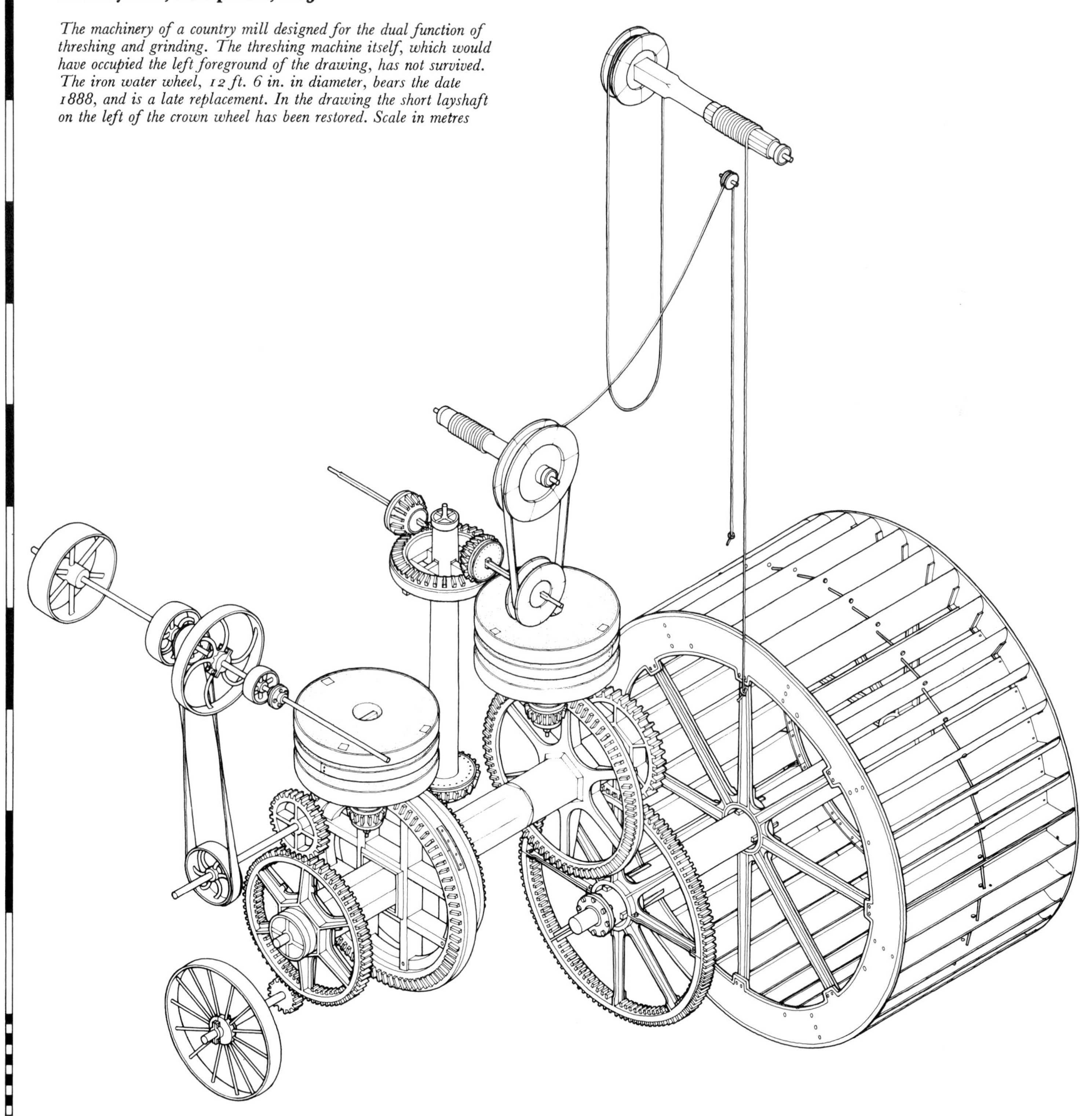

RIGHT
Hockley Mill—interior of the ground floor, showing the horizontal shaft (top left) and the pit wheel

OPPOSITE
Hockley Mill—wooden gear wheels on the main shaft: note the hardwood dowels driven obliquely through shanks of cogs. The wallower and the foot of the vertical shaft can be seen top right

BELOW
Hockley Mill—the stone floor: note the crook string passing down through the floor

Hockley Mill—ground floor, showing the tentering screw (centre left), crook string and twist peg (centre right) and stone nut (top right)

Like the farms and barns of the countryside, mills were soundly built of the materials locally available. The mill drawn by the artist of the *Luttrell Psalter* early in the fourteenth century was a small timber framed building with walls of wattle and daub, and a roof of thatch. Only the water wheel, turning against the gable end, distinguished it from the peasant's cottage, and other early representations of watermills indicate that very similar buildings were in use throughout Europe at that time. The fact that few ancient mills survive is not surprising when consideration is given to the variety of abnormal hazards to which they were exposed. Many stood of necessity on marshy ground, which provided a poor bearing for foundations. Timber piles rotted away in the waterlogged soil, and rivers in flood presented a recurring danger. The vibration of machinery, and the imposition of heavy intermittent loading tended to weaken the structure, while friction between the stones might at any time spark off a fire which could spread with explosive force through the dust-laden air. For these reasons the life-span of the building seldom exceeded two centuries, and old mills display bulging walls tied in with iron straps, and sagging floors, as the honourable scars of long service.

The small country mill was built in the vernacular style of the district, and might thus be of stone, brickwork, or half timbering. During the eighteenth and nineteenth centuries, many large timber-framed mills were erected, and these were commonly clad with weather-boarding, a material which became particularly associated with mill work. In Britain boarding was either tarred or painted white, and where fire and decay have spared them these gaunt wooden buildings stand as notable landmarks. It will be these soaring towers of painted boarding, dominating the landscape for miles around, which will remain for many the most characteristic of all watermills.

Medieval watermill from the Luttrell Psalter, *early fourteenth century: note the overshot wheel and eel traps*

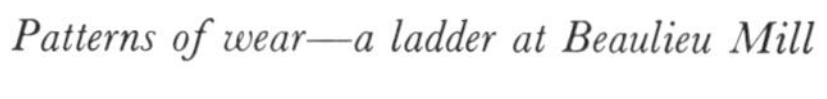

Patterns of wear—a ladder at Beaulieu Mill

Wickhambreux Mill, Kent

Brickwork and half-timber at Sherfield on Loddon, Hampshire

RIGHT
Commemorative tablet at Nursling Mill, Hampshire: the inscription reads, 'This building stands on a frame of large beech timber which was given by Sir Richard Mill, Bart. in memory of whose kindness this stone was placed here by T.C.K. 1728'

BELOW RIGHT
Brick and tile at the Chesapeake Mill, Wickham, Hampshire. The floor beams of the mill are of timber from the American frigate Chesapeake, *captured off Boston by* H.M.S. Shannon *in 1813*

Mills were usually treated as strictly utilitarian structures, and little effort was made to add architectural embellishment. Decoration was restricted to the decent minimum required by the custom of the period, and might consist of no more than a symmetrical pattern of half timbering, or later, a dentil course under the eaves. Sometimes a flourish of ornament was introduced in a carved date stone over the door, and sheaves of corn might supply the theme for a decorative cartouche or finial. But in general, mills share the qualities associated with other working buildings in the countryside, and give pleasure to the eye through their functional design and unselfconscious use of traditional materials. Occasionally they were re-built, or re-faced, in a deliberately picturesque style to serve as features in a contrived landscape. Behind Gothic battlements or Classical portico the mill would work on as before; a rare example of a purely decorative mill survives at Versailles, in the sad little building erected as a plaything for Marie Antoinette.

Some few richly ornamental buildings stand as exceptions to the general rule. Bourne mill near Colchester, with its wealth of Jacobean detail, evokes most strongly the magical quality of that brief period of English life which was to end abruptly with the Civil War. Large numbers of mills were owned by the Church, and surviving remains suggest that these were often buildings of some consequence. West Harnham mill near Salisbury preserves good stone detailing of Tudor date, but may perhaps have been designed originally for some purpose other than milling.

Many British mills were distinguished by the use of the mansard roof, a continental form never generally adopted for buildings of other types. The double pitch had obvious advantages for mill work, giving increased headroom above the bins. But the most characteristic architectural feature was undoubtedly the 'locum', a boldly projecting bay through which sacks could be hoisted in a single vertical lift, from the wagon up to the level of the bin floor. These great over-sailing dormers, supported on massive timber gallows brackets, often add a dramatic point of emphasis to an otherwise severely regular elevation. While small mills like that at Abbotstone might not boast a 'locum', loftier buildings invariably made use of this arrangement, until the introduction of the elevator rendered it no longer necessary.

LEFT
Marie Antoinette's Mill, Versailles

BELOW
Ovington Mill, Hampshire

RIGHT
Classical frontispiece to the Abbey Mill, Winchester

OPPOSITE
Bourne Mill, Colchester, Essex: built in 1591 on a mill site dating from the twelfth century. The overshot wheel, 26 ft. in diameter, is housed within the building

BELOW
West Harnham Mill, Salisbury

THE MILL HOUSE

The miller's own living accommodation reflected his changing social status over the years. On the medieval manor his position was a humble one, and in Chaucer's time he lived with his family in the mill. As the rigid restrictions of feudal custom were slowly relaxed, first in the towns and later in the countryside, the miller became a more prosperous member of the community. He occupied a mill house, which often formed an extension of the mill itself, sharing the same roof. Many buildings of this type survive, subtle changes in detailing distinguishing the workaday mill from the more genteel mill house adjoining. In the eighteenth and nineteenth centuries the miller was frequently an important figure in the locality, and many handsome Georgian mill houses were built. These often stood detached from the mill, and with their sash windows and fanlights they vied in elegance with the houses of the minor gentry.

RIGHT
Ickham Mill, Kent

LEFT
Elstead Mill, Surrey

BELOW
Chilham Mill, Kent

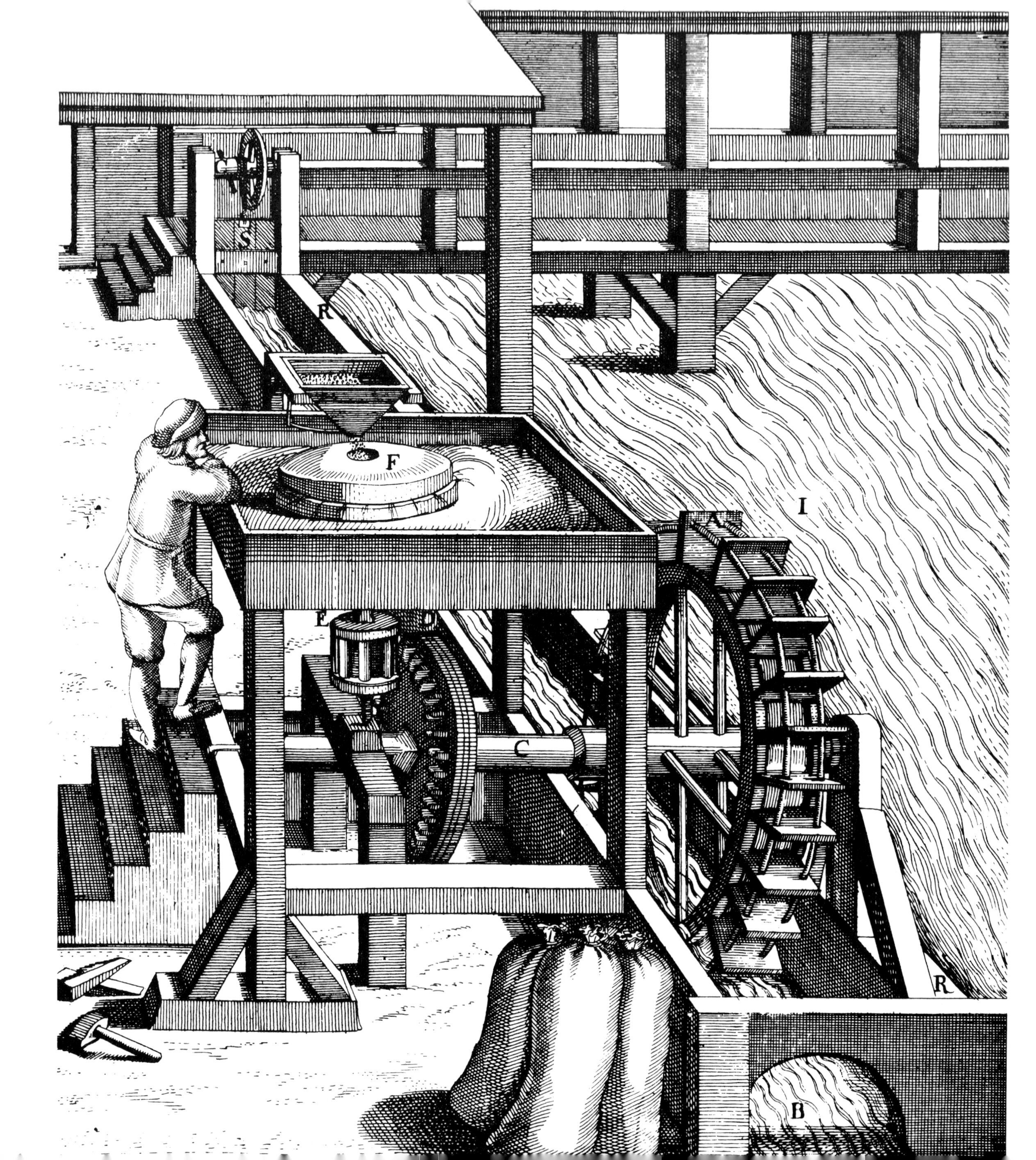
S
R
F
I
E
C
R
B

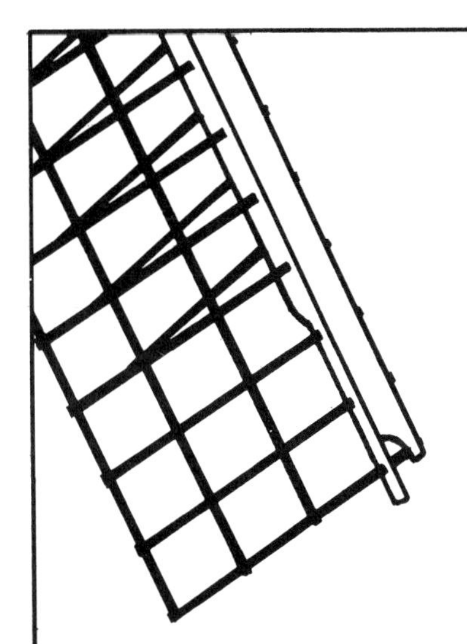

3 The Machinery of the Watermill

Throughout the Middle Ages the machinery of the mill was of timber, skilfully framed together and fastened with wooden dowels or trenails. Wrought iron was a costly commodity and its use was strictly limited. Even in the sixteenth and seventeenth centuries, when engineering textbooks were beginning to appear in Europe, the use of iron was reserved for the gudgeons on which timber shafts and axles revolved, and for similar parts subjected to exceptional wear or loading. The tradition of timber construction never died out, although cast iron was used to an increasing degree from the eighteenth century onwards. Timber was in fact a very suitable material for this slow-running machinery, and when one stands in a working mill one cannot fail to be impressed by the smooth and quiet meshing of the wooden gears.

In Britain oak was customarily used for wheels and shafts, and holly for trenails. Cogs were shaped from apple wood, or from some other native hardwood such as hornbeam or beech. Elm was favoured for the floats of water wheels, as it was for coffins, because of its resistance to rot, while stone was frequently employed for bearings. In old mills the machinery is seldom of a single period. Parts were replaced when they became worn, and the water wheel, always liable to damage, and subjected to constant immersion, is frequently of a later date than the remainder of the machinery. The undershot wheel of timber at Beaulieu mill in Hampshire is therefore a rare survival. It is of primitive 'compass arm' construction, with radial spokes mortised to a heavy wooden axle and secured with wedges. Iron is used only for fastenings, and for tie rods linking the floats. Each rim, like that of a wagon wheel, is built up from segments or 'felloes' of oak, but the design precludes the use of an encircling iron tyre, the strength and symmetry of the wheel depending upon the craftsmanship of the millwright.

Both undershot and overshot wheels were known and used during the Middle Ages. The typical early mill was no doubt powered by a simple undershot wheel, not unlike the example shown on p 39. But illuminated manuscripts of the fourteenth century show that overshot wheels, served by timber launders, were installed where conditions were suitable. In such wheels the floats were enclosed between deep wooden rims in order to retain the water. It was later realised that even a relatively small change of level, induced by a natural or artificial

A corn mill with undershot wheel from Theatrum Machinarum Novum *by G. A. Böckler, c. 1662*

A cross-head gudgeon removed from a timber shaft

The tide mill at Beaulieu

Beaulieu Tide Mill, Hampshire

An undershot timber wheel and axle, of traditional form, driving two pairs of millstones. The iron hursting, of late nineteenth century date, is a replacement, and incorporates tentering mechanism controlled by handwheels. Beaulieu Mill retains an old manual hoist, mounted below the ridge of the roof. The drawing shows the water wheel restored, and the position of the sack hoist adjusted. Scale in metres

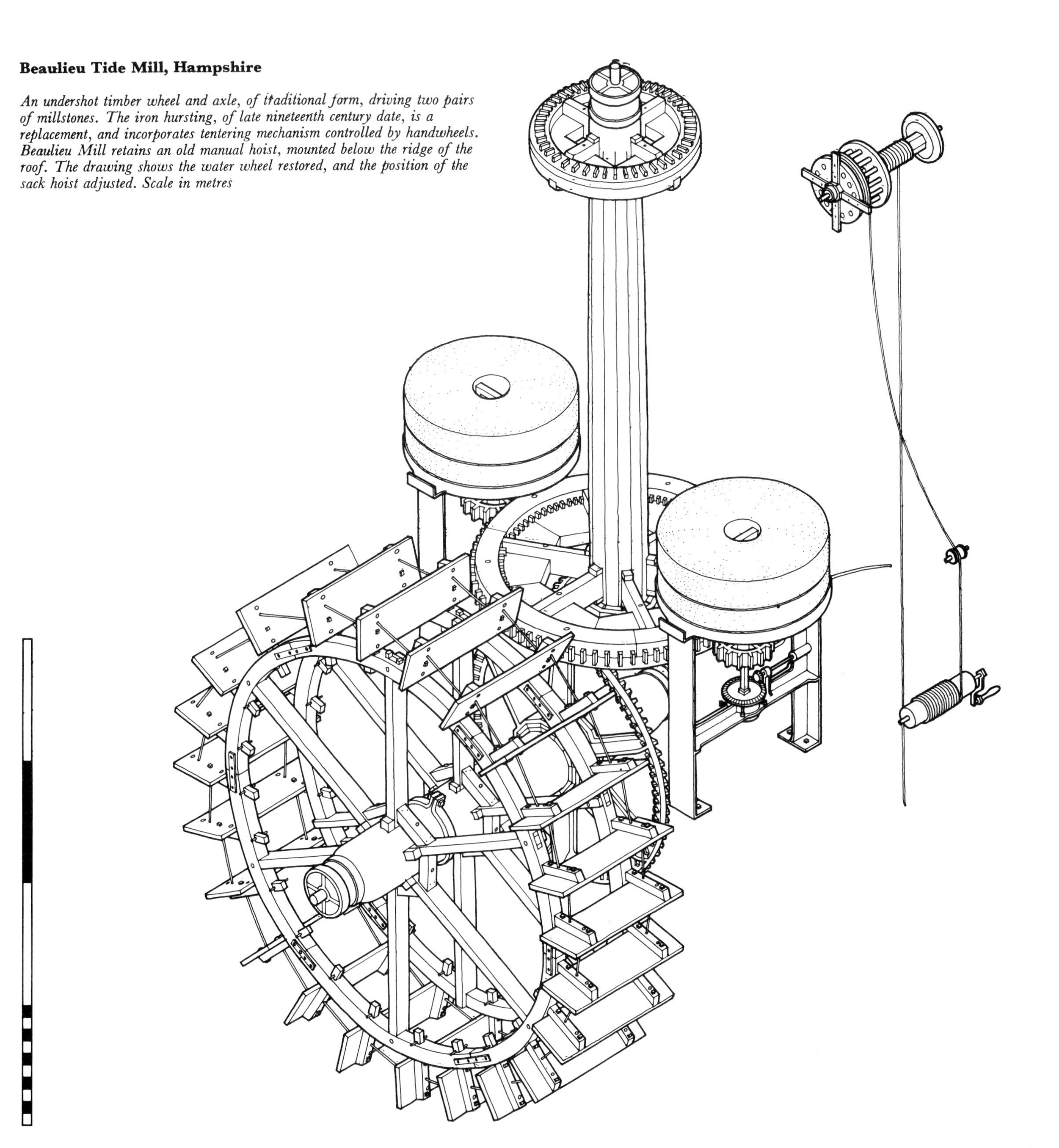

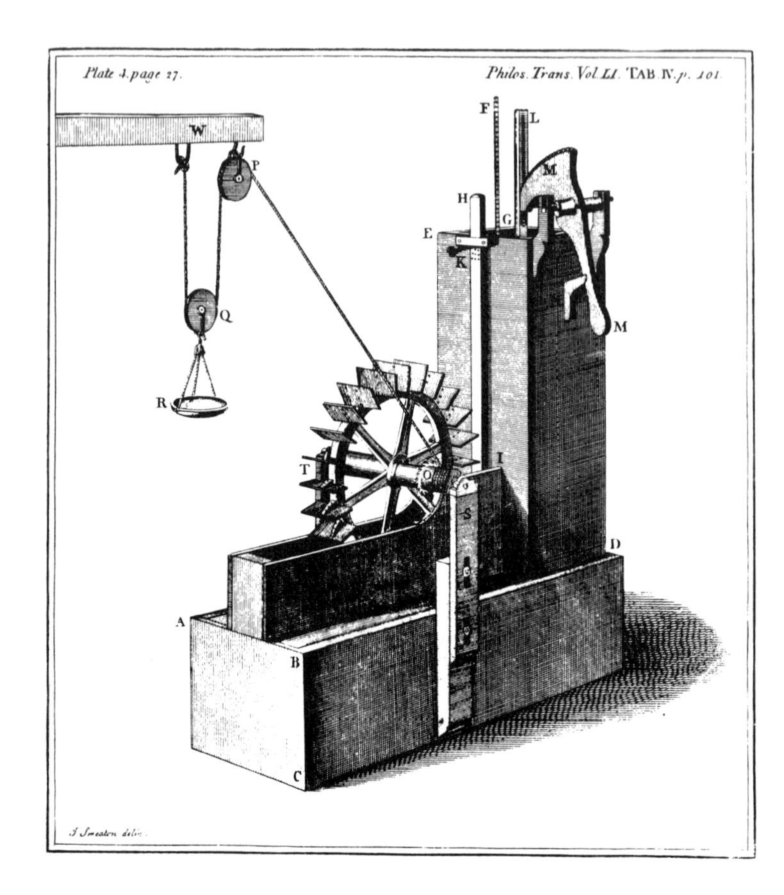

A model used by Smeaton for research into the design of water wheels

BELOW
Hand mill dated 1745, from an Oxfordshire farm, with wooden gearing following the medieval tradition

weir, could materially increase the power of the mill, and by the sixteenth century a third basic type, the breast-shot wheel, had made its appearance. Authorities such as Sir Anthony Fitzherbert, whose *Boke of Surveying* was published in 1538, extolled the virtues of 'braste' and overshot wheels, but more than 200 years were to elapse before their superiority was scientifically established by John Smeaton's careful experiments.

A serious disadvantage of the 'compass arm' wheel lay in the fact that the deep mortices weakened the axle, and at the same time allowed water to penetrate to the heart of the timber. Many millwrights therefore preferred the alternative 'clasp arm' method of construction, using pairs of parallel spokes halved together at right angles, to enclose the axle, which was subsequently centred and secured in position by means of wedges. Inside the mill the drive was transmitted from the horizontal axle to a vertical shaft, or spindle, by two wooden gears. The larger of these, mounted on the axle, was a face wheel fitted with hardwood pegs in place of the cogs of later years. Known as the 'pit wheel', since its lower half was normally accommodated in a pit sunk in the floor of the mill, it engaged with the 'wallower' mounted near the foot of the vertical shaft. The 'wallower' was either a smaller pegged face wheel, or more often a lantern pinion, a simple type of gear wheel much used by the medieval millwright, consisting of a ring of staves mounted between two hardwood discs. In Britain it was eventually superseded by the more efficient cogged pinion, but examples may still be seen at work in continental mills. Where the mill stream ran under the building, the

A lantern pinion from a Dutch mill

gearing was often simplified by the omission of the 'pit wheel', the lantern pinion being driven directly by pegs projecting from the rim of the water wheel.

In constructing this wooden machinery, the millwright was concerned not only to achieve strength but also to counter the natural movement of the timber which formed his raw material. More often than not this was oak—'too strong to lie quiet'—and its strength lay in the direction of the grain. Wheels were laminated, the two thicknesses being pinned together with trenails and arranged so that the grain crossed at right angles. When the diameter exceeded four feet, a rigid clasp arm frame was commonly employed to support the rim, which was built up from segments of timber, bolted and dowelled together with staggered joints. Cogs were roughly shaped on the bench, then fitted to the wheel and finished in position. Where possible the mortises passed right through the rim, and the shanks were secured in place by wedges or pins. The millwright avoided the use of simple gear ratios by the insertion of a 'hunting cog'. This precaution ensured that two wheels would mesh together differently at each revolution, thus preventing uneven wear. But no amount of use would make ill-fitting gears run smoothly, and the greatest accuracy was necessary in the initial setting-out. The machinery of the mill was strictly functional, a minimum of decoration being provided by the cleanly cut chamfers on spokes and shafts. Where years of use have not imposed their own patterns of wear, the firm strokes of the millwright's chisel may still be seen, and one cannot fail to be impressed by the quality of the workmanship in these great wooden engines.

The medieval corn mill, like its Roman forerunner, was equipped with a single pair of stones, but the traditional arrangement was later modified by the introduction of a main vertical shaft, capable of driving two or more pairs through spur gearing. This new layout, which may well have originated in the windmill, constituted a major advance. It was widely adopted during the latter part of the sixteenth century, and was to remain the accepted method of power transmission until the last watermills were built.

From the middle of the eighteenth century onwards cast iron was employed to an increasing degree in mill work. There were obvious advantages in using iron for parts subjected to immersion, and various techniques were adopted for the construction of water wheels. Because large castings presented difficulties, many composite wheels were installed, and at first iron was reserved for hubs and rims, the latter built up from segments bolted together. By the end of the century, however, many fine wheels were being produced with complete frames of iron, timber being retained only for the floats, and the short projecting 'starts' to which they were bolted. The delicate skeletons of these great wheels may still be seen in many deserted mills. With narrow socketted rims and tapering cruciform spokes, they are notable for the elegance of their design.

At this period the principles of water power were being investigated in an increasingly scientific manner, and advances were made in the design of the three basic wheel types. A marked improvement in the efficiency of the breast-shot wheel was achieved by the introduction of a close-fitting curved facing to the weir, while the primitive undershot wheel was given a new lease of life by the brilliant French engineer, J. V. Poncelet. In 1824 he introduced the design which was to bear his name; an undershot wheel of dramatically increased potential, capable of extracting the maximum amount of energy from a fluctuating mill stream. The velocity of the water striking the wheel was increased by channelling it below an inclined sluice, the curvature of the buckets being carefully calculated to allow water to enter and leave the wheel without shock. For the first time undershot wheels were able to compete in terms of power output with rival types, and large numbers of Poncelet wheels were installed during the nineteenth century. They were constructed entirely of iron, sheet metal being peculiarly suitable for the subtle shaping of the buckets.

Both breast-shot and overshot wheels benefited from the introduction of 'ventilating buckets', while the problem of 'back-watering', which could reduce the power of the conventional overshot wheel, was countered by the introduction of the 'pitch-back' type. 'Back-watering' occurred when the lowest floats became submerged in the race. By modifying the launder, however, it was possible to apply power to the opposite side of the wheel, thus reversing the direction of rotation and setting the floats moving in the same direction as the tail water. The power of the mill was thus increased rather than diminished, and the 'pitch-back' wheel was frequently utilised where marked fluctuations in water level were encountered.

Other components for which cast iron quickly came into general use were the pit wheel and wallower. The former could often be partly submerged, and in such conditions the superiority of iron was indisputable. The ancient combination of face wheel and pinion gave place to the more efficient bevel gearing, the pit wheel being cast with a slotted rim, and fitted with hardwood cogs. It was usual to cast the wallower completely in iron. Hardwood and metal gears meshed together happily and quietly, but the opposition of iron to iron was avoided by millwrights wherever possible. Large overshot wheels presented certain special

Iron breast-shot wheels at Titchfield Mill, Hampshire

Watermills of the late sixteenth century from Künstliche Abrisse Allerhand Wasser-Wind-Ross- und Handt Muhlen *by Jan de Strada (engraving by Phillip Galle), 1617.*
In early mills, each water wheel served a single pair of stones. Here the simplest form is illustrated, the lantern pinion meshing with pegs projecting from the rim of the wheel

LEFT
The principal gearing at Abbotstone Mill, Hampshire, viewed through the front of the hursting. In the centre can be seen the foot of the main shaft

BELOW LEFT
A pitch-back wheel of composite construction at Skeagh Mill, Sticklepath, Devon. Water, delivered from the launder (top left), was discharged over the back of the wheel

BELOW
Abbotstone Mill—a view through the side of the hursting, showing the pit wheel and wallower (bottom right) with the stone nut and great spur wheel above. Grease from the thrust bearing of the stone spindle runs down the face of the bridge tree

Runner stones standing against the wall of a mill. The stone on the right is a French burr, showing the characteristic 'patchwork' construction

problems. It was necessary for them to be heavily built in order to withstand the stresses imposed by the traditional arrangement of pit wheel and wallower. During the latter part of the nineteenth century this problem was overcome by fitting a toothed driving ring to the rim of the water wheel itself, engaging with an external pinion which carried power into the mill. By this means it was possible to effect a considerable reduction in weight, with a consequent increase in power.

Iron construction was adopted but slowly for the remaining machinery of the mill. Timber was still usually employed for the main vertical shaft which transmitted power through the height of the building, and in many surviving examples the great spur wheel is also of timber. The small pinions through which the millstones were driven were originally of the lantern type, but these were replaced by solid wooden wheels, bound top and bottom with iron, and mortised for hardwood cogs. Later these in turn gave place to slotted iron pinions, cogged in hardwood. To allow either pair of stones to be taken out of gear, it was usual to make several adjacent cogs removable. These were secured in place with iron pins, and were known as 'slip cogs'. In the small country mill the main shaft was carried up to terminate at a bearing bolted to one of the tie beams, and near its head was mounted a crown wheel which again might be of timber or iron. The crown wheel supplied power to the sack hoist, and, through lay shafts, to the auxiliary machinery of the mill, which was to grow in bulk and importance with the passage of time.

At the heart of the corn mill were the stones themselves. They were arranged in pairs, the upper 'runner stone' rotating above the lower fixed 'bed stone' or 'ligger'. Sizes varied over the years, but a diameter of about four feet came to be generally accepted as most satisfactory. Such a stone could weigh over a ton when new, and functioned most efficiently at a speed of 125–150 revolutions per minute. Sources of suitable material were few and far between and millstones of good quality were very valuable. The best were produced at certain continental quarries, and even during the Roman period millstones from the Rhine valley were being imported into Britain. This trade was to continue during the Middle Ages, but millstones were heavy, transport was slow and costly, and many millers were forced to make use of the best stone locally available. In Britain this was millstone grit, a hard sandstone quarried in the peak district of Derbyshire, but many less satisfactory stones were also employed, including sandstone from the west Midlands and granite from Wales and Dartmoor. 'Dutch blue' and 'Cullin' stones, the latter taking their name from the city of Cologne, were shipped to Britain from the continent and were widely used over the years. Most highly prized of all were 'French burrs', skilfully pieced together from blocks of a particular type of quartz, quarried near Paris. This material occurred in small deposits only and it was necessary to shape and match the pieces with great care. They were then jointed in cement, bound with iron hoops, and backed with plaster of Paris. Pairs of French 'burrs' survive in many English mills, where they are easily identifiable by their patchwork appearance. In America, millstones were quarried on Mount Tom in Connecticut, and later in Ohio. Other local stones of varying quality were also employed, but many mills used European millstones, and these were imported in considerable numbers.

The working surfaces of the stones required careful preparation, and skilled stone dressers were craftsmen in their own right. When dealing with a new stone it was necessary first of all to produce a perfectly smooth surface. A laminated wooden straight edge was used with 'raddle', a composition of red oxide, to detect the high spots, and these were rubbed down with a fragment of burr stone. An extraordinary degree of accuracy was required, and achieved, in this operation. The dresser tested the edge of his wooden staff from time to time against a proof staff of cast iron, which was kept carefully protected from damage in a wooden case. The final aim was to achieve a surface very slightly dished, or hollowed, towards the central eye, ready to receive the system of 'furrows' which performed the actual operation of grinding. These were set out in patterns established through years of trial and error to suit the particular use to which the stone was to be put. In common dressing, the surface was marked out into ten equal 'harps' or sectors, divided not by true radii, but by lines drawn at a tangent to the eye of the stone. This basic method of division, to be seen on Roman millstones, was to remain in use until the end of the traditional corn mill. Each 'harp' was subdivided into alternate 'lands' and 'furrows', the latter cut to a depth which varied between one half and three quarters of an inch, and finished with a sharp arris on one side, an even slope up to the 'land' on the other. Finally a system of fine parallel grooves, the 'stitching' or 'cracking', was added to the surface of each land, a process calling for both skill and precision, experienced men being able to cut as many as 16 'cracks' to the inch. The dressing was so arranged that when the stones were in their working position, face to face, the 'furrows' crossed at each revolution, cutting the grain with the action of scissor blades. As the stones became dulled with wear they ceased to grind efficiently, and frequent re-dressing was required. This involved the deepening of 'furrows' and the renewal of 'stitching', and with a pair of French burrs, might become necessary every two or three weeks.

Dressing was performed with the 'bill', a cutting tool resembling a double-ended wedge forged from high carbon steel, and the pointed 'pick'. These tools, mounted in turned wooden handles or 'thrifts', were used with a pecking action to chip away the surface of the stone. A stock of 'bills' was always maintained, as one man would blunt three or four in the course of an hour's work, and a grindstone therefore formed an essential part of the equipment of every mill. From time to time, 'bills' were re-ground and tempered at the smithy, until they became too light for further use. A few blacksmiths were able to scarf

Mill bills and thrifts, the tools of the stone dresser. In the background are wedges used in raising the runner stone, in the foreground a wooden trammel

Stone dressing in a Dutch mill: note the bist, or cushion, on the left

Stone dressing at Milton Mill, Milnthorpe, Westmorland. The miller is working on the bed stone; the runner has been raised, and stands on the right

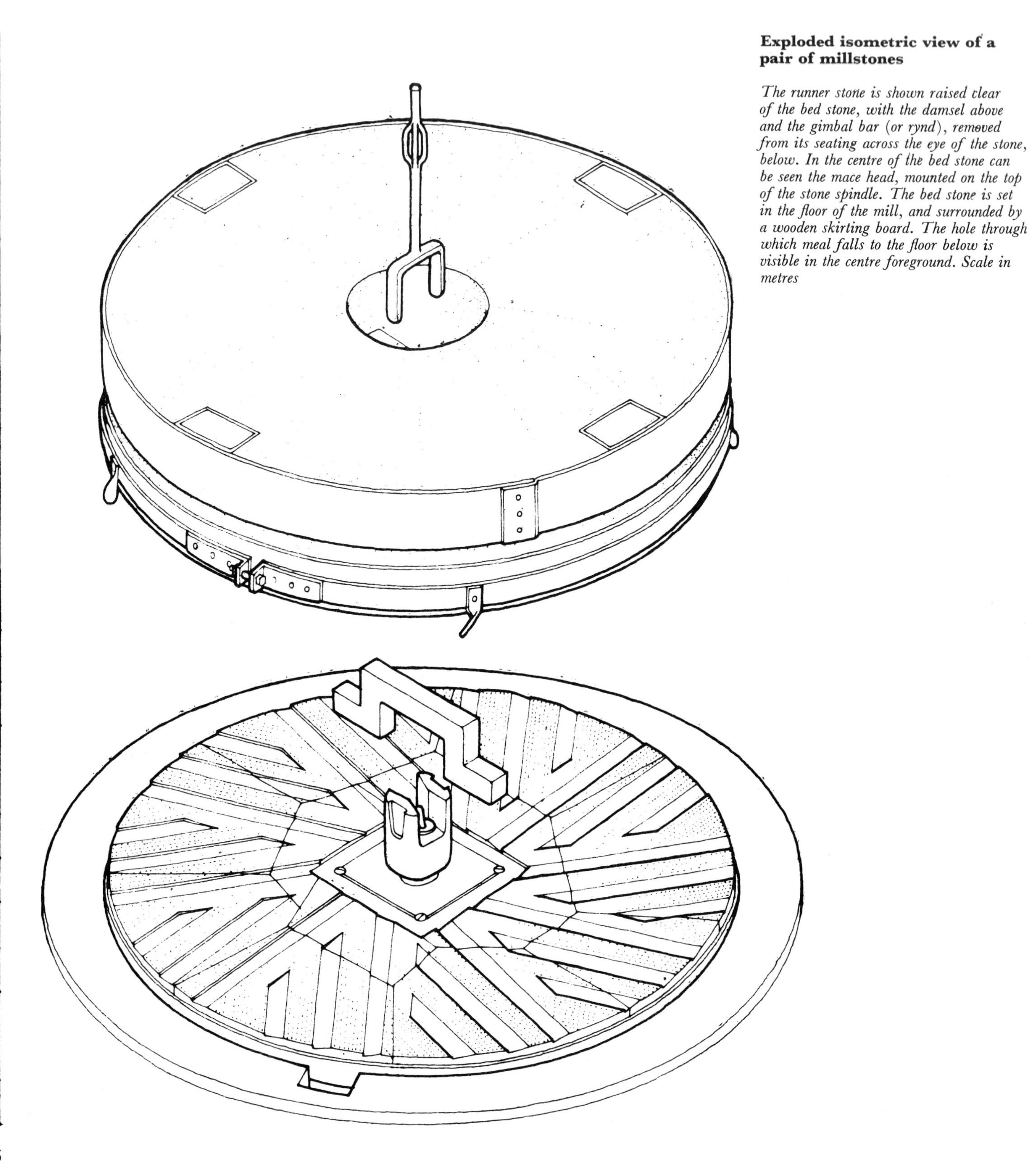

Exploded isometric view of a pair of millstones

The runner stone is shown raised clear of the bed stone, with the damsel above and the gimbal bar (or rynd), removed from its seating across the eye of the stone, below. In the centre of the bed stone can be seen the mace head, mounted on the top of the stone spindle. The bed stone is set in the floor of the mill, and surrounded by a wooden skirting board. The hole through which meal falls to the floor below is visible in the centre foreground. Scale in metres

A stone fitted with Messrs. Clark and Dunham's adjustable balance weights

BELOW
A pair of stones at Hockley Mill, Hampshire: note the cavities for balance weights in the top of the runner stone. Meal emerging from between the stones was swept along by the projecting tag to fall through the rectangular opening (bottom right)

weld a pair of worn 'bills' together, and restore the temper of the steel, and such men earned themselves a considerable local reputation among millers. But so personal was this knack that at least one Hampshire smith was unable to pass it on to his son, himself a competent craftsman. With the decline of the working mill, many millers were forced to undertake the dressing of their own stones, but in the nineteenth century pairs of itinerant stone dressers made a living by tramping the country from mill to mill. Such men's hands became discoloured by minute particles of steel struck from the 'bills', which lodged under the skin. Before engaging a man whom he did not recognise, the miller would ask him to 'show his metal', and judge from the state of his hands whether he was indeed an experienced craftsman.

Before the runner stone was placed in position, the bearings of the spindle on which it was to turn were carefully adjusted to allow the stone to rotate in a perfectly horizontal plane. This operation of 'brigging the spindle' was greatly facilitated by the introduction of the iron 'bridging box', a square container in which the position of the thrust bearing could be altered by means of projecting hackle screws. The actual clearance between the grinding surfaces of the stones required very fine adjustment, in order to produce meal of the desired texture, taking into account both the hardness of the grain and the speed of the mill. This process was known as 'tentering', and was effected by raising or lowering the ends of the stout timber beams, or bridge-trees, which carried the stone spindles.

The bed stone could be wedged up from below until it lay in a perfectly horizontal plane, but it was necessary to balance the runner stone exactly. This was done by running lead into cavities cut in the back of the stone, or later by the use of patent adjustable balance weights. These precautions were necessary to ensure that the stones never actually touched when in motion. Friction could produce more wear in minutes than weeks of grinding, and a shower of flying sparks that could easily fire the mill. With a ponderous stone spinning at high speed spectacular accidents could occur; it was not unknown for a small fragment of scrap metal, hidden in the grain, to throw the runner stone off its spindle and through the floor of the mill. It was essential to keep the stones supplied with grain, and alarm bells were fitted to give

Isometric view of the vat or tun enclosing a pair of millstones

The rectangular opening in the top of the vat is spanned by a wooden framework, or horse, supporting the hopper (top left) from which grain falls to an inclined spout or 'shoe' and thence to the eye of the runner stone. The shoe, partly visible through the framing of the horse, is shown in greater detail below, resting beside the vat. When in use its angle is altered by the adjustment of the crook string, here passing down through a hole in the floor boards (right). The shoe is tensioned against the revolving damsel (centre) by a cord attached to a wooden spring, the 'miller's willow', shown stapled to the top of the vat. Scale in metres

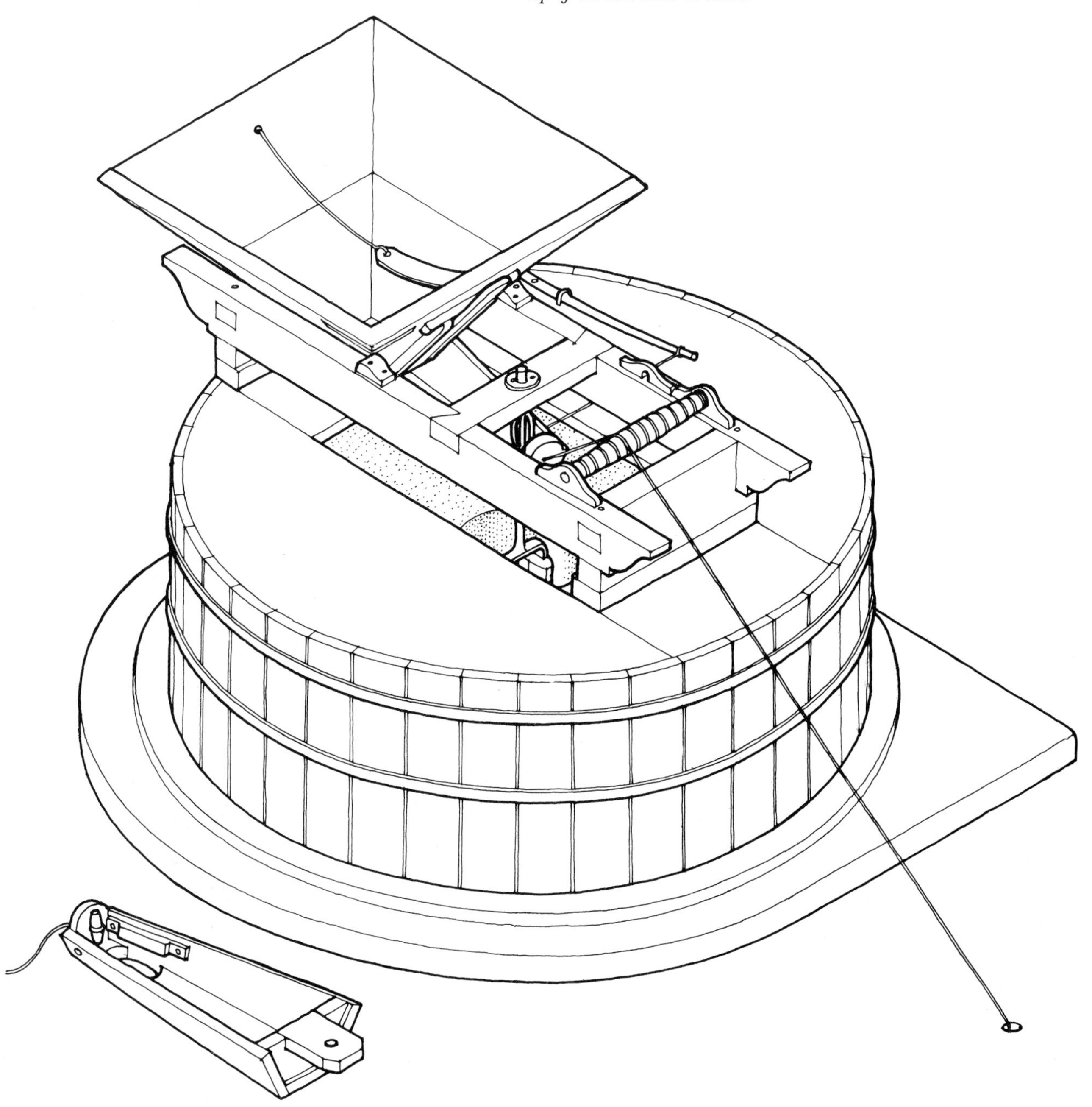

warning when the hoppers became empty. One of the miller's children would sometimes be stationed by the bins, to see that the grain fell steadily down to the stones below. The miller could tell when all was well by the 'singing of the stones', but in a watermill the roar of the race tended to drown other sounds, especially where the wheel was housed within the building. At work or at rest mills were full of the sound of running water.

The stones were enclosed in a light wooden 'tun' or 'vat', across the top of which lay a framework known as the 'horse'. This supported a square wooden hopper and a tapering spout, or 'shoe', which fed grain into the eye of the runner stone. A forked iron rod, the 'damsel', passed through both 'shoe' and 'horse' to engage with the revolving rynd below. As the 'damsel' turned, it rattled against the 'shoe', shaking grain down into the stones. When the mill was at work the damsel was never silent, hence its name. As the stones gathered speed, so the vibration increased and more grain was discharged. Manual adjustment could be made with the 'crook string', which altered the angle of the 'shoe', and was controlled by a twist peg on the floor below. To ensure that the 'shoe' was smartly rapped by the turning 'damsel', it was tensioned by a simple spring. In late mills the spring was sometimes of steel, but more commonly it took the traditional form of the 'miller's willow', a green wand of hazel or ash stapled to the top of the 'vat'. In early times an even more primitive method had been employed to agitate the 'shoe', a block of stone, the 'clapper', being allowed to drag across the uneven surface of the spinning runner stone.

The actual operation of grinding had been for many years the only mechanical process performed in the corn mill. Sacks of grain and meal were carried on the miller's shoulders and if the need for a sack hoist was felt, it was no doubt met by a rope passed through a block hung from a convenient beam. Later it became customary to install a form of windlass, the hoist rope being wound on to a horizontal shaft or barrel. The load was raised by hauling down on a second rope, or endless whip, and mechanical advantage was obtained by winding this line in the reverse direction around a pulley wheel of larger diameter. A few primitive devices of this sort may still be seen, and an interesting example survives at Beaulieu, but they were generally superseded by the power-driven sack hoist, the first of many secondary machines to be installed in the mill.

The stone floor at Ecchinswell Mill, Hampshire. The main shaft (centre) is flanked by two pairs of stones: on the left, the vat, horse and hopper are in place. On the right the vat has been removed and the runner stone is being raised.

The mechanical hoist may well have originated in the windmill, where the location of the main gearing above the level of the stones made it relatively easy to contrive some form of simple friction drive. The general introduction of a main vertical shaft made power similarly available at high level in the watermill, enabling various forms of hoist to be developed. In most cases a horizontal winding drum or barrel was mounted high up under the ridge, and connected by a slack belt to a lay shaft driven from the crown wheel below. The mounting was so contrived that a steady pull on a control rope would lift one end of the barrel, thus tensioning the belt and setting the hoist in motion. Belt drive was by no means universal, but certain features were common to most mills. The hoist chain terminated in a ring, so that a running noose could be slipped over the neck of the sack. A pair of trap doors at each floor level opened upwards on leather hinges as the sack passed through, and fell to behind it. Hoists generally drove in one direction only, and the miller had to overhaul the chain by hand ready for the next load. It was also necessary for him to manhandle the sack along the top floor of the mill, and empty it into the storage bin selected. Thus while the hoist eased the miller's burden, it still left a great deal of heavy manual labour undone, and if one man was working the mill on his own, he had no alternative but to climb to the bin floor after sending up each sack.

During the latter half of the eighteenth century, a young American millwright, Oliver Evans, set himself the task of reducing this great volume of unproductive work. Born in Newport, Delaware, in 1755, Evans was apprenticed to a wheelwright. He soon showed himself to be not only an inventive genius but also a practical engineer, and before he was 30 years old he had equipped a working mill with a comprehensive range of labour-saving devices. These were principally concerned with the carriage of grain and meal within the building, and included both the elevator and the auger. The former consisted of an endless belt, equipped with either buckets or flat rungs. By its use loose grain, delivered in bulk to the ground floor of the mill, or even brought alongside in barges, could be raised without effort to the top of the building. An added advantage lay in the fact that it placed a light and continuous load on the machinery of the mill, in place of the heavy intermittent loading imposed by the sack hoist. Another of Evans' inventions was the 'descender', which allowed material to drop at a controlled speed under its own weight, and in general the fullest use was made of gravity in the progressive passage of grain and meal downwards through the building during the milling process. The horizontal movement of grain was effected by conveyors or creepers, narrow wooden troughs containing rotating augers. Like many of Evans' innovations the creeper was not an original invention, but rather an inspired application of an existing principle.

These new machines saved the miller much time and labour, and the concept of mechanisation was widely adopted in America, where the growth of the industry was affected by very different factors from those which had obtained in medieval Europe. The opening up of the Middle West brought vast tracts of prairie land under the plough, producing grain far in excess of local needs. This type of farming demanded an efficient transport system, and this in turn made possible the establishment of huge milling centres, strategically sited to take advantage of the enormous reserves of water power available. In the early days of colonial development New York had enjoyed a virtual monopoly of merchant milling, but by the middle of the eighteenth century the lead had passed to the region of Chesapeake Bay, where first Philadelphia and later Baltimore were able to exploit the rivers running down from the Appalachian Mountains. These cities were in turn surpassed by

FAR LEFT
Beaulieu Mill—the old sack hoist

LEFT
The sack hoist at Ecchinswell Mill. The driving belt is tensioned by raising the end of the bollard

BELOW LEFT
Sack scales at Beaulieu

BELOW
An automatic mill from the The Young Millwrights Guide *by Oliver Evans* c. *1785. This illustration shows elevators (5, 24 and 39), conveyors or creepers (15 and 45) and the hopper boy (25).*

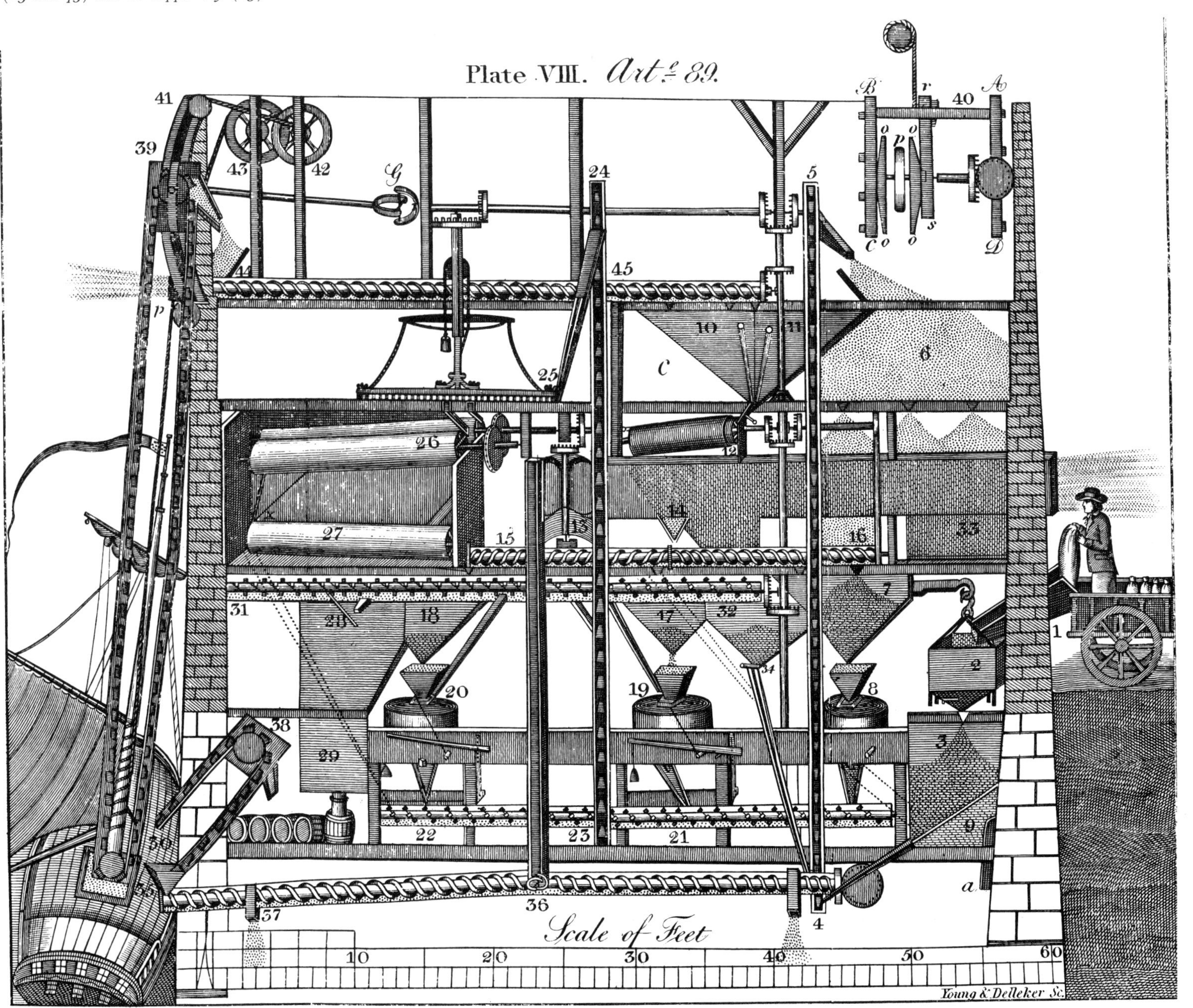

LEFT
A corn mill with overshot wheel, c. 1662, Böckler: note the early form of bolter

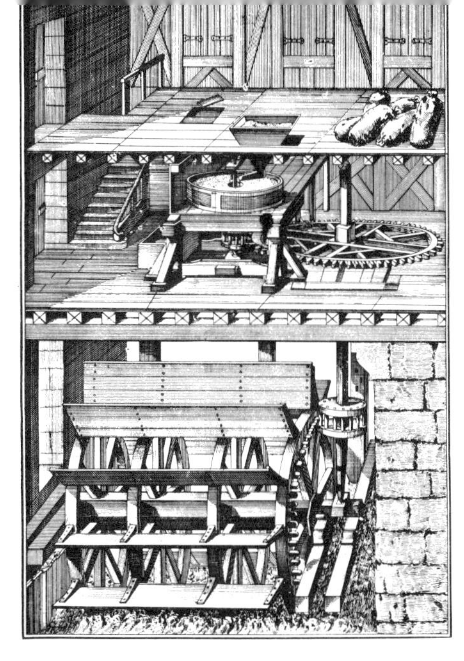

FAR LEFT
The wire machine at Bembridge windmill

LEFT
An American watermill, c. 1795: note the lantern pinion driven off the rim of the wheel

Minneapolis, which had become, by the late nineteenth century, the world's largest centre for commercial milling. Competition here was fierce, and Evans' improvements found ready acceptance. They were adopted more slowly in Europe, where milling still tended to follow the traditional pattern. The large modern mills set up at major seaports to deal with imported grain embodied the new machinery, but many country mills were to rely on the primitive sack hoist until the end of their days.

If flour was to be sifted in the medieval mill it was shaken in a hand sieve or 'temse'. The words, 'He won't set the temse on fire', originated in the mill, and were no doubt first spoken of some luckless miller's boy. In fact the miller only sifted flour for his own use, or at the special request of one of his customers, this process being regarded as a part of the baker's duty. It was not until the year 1502 that the first mechanical sieve was introduced to perform this task. According to tradition it was invented by Nicholas Boller, an Austrian, and in succeeding centuries the process of 'bolting' was to grow steadily in importance. An illustration in Ramelli's treatise on machines, published in 1588, shows a cylindrical 'bolting' bag of cloth arranged below the stones to receive the freshly ground meal. The bag was shaken by a wooden rod, the end of which engaged with the staves of the revolving wallower. Whilst primitive 'bolters' of this type survived in use in Germany into the nineteenth century, many more sophisticated machines were devised in England and America to cope with the increasing demand for finely graded flour. With the passage of time these were to become steadily more complicated and bulky, until they occupied a large proportion of the space available within the mill.

The type of bolter which came to be installed in many country mills in Britain consisted of a cylindrical frame rotating on an inclined axis and enclosed in a wooden casing. Over the frame was fitted a tubular bolting cloth, reinforced with leather bands and fastened with draw strings. Meal entered the revolving cylinder at its upper end, and weighed down the cloth until it brushed against a series of longitudinal bars, which shook the flour through the woven fabric. Early bolters were turned by hand, but methods of transmitting power from the main shaft of the mill, either by gearing or by pulley and belt, were soon devised. Bolting cloths were at first made from wool, and those manufactured in England achieved a high reputation on the continent. In later years, woollen fabric was superseded, first by calico and finally by silk, and bolters were in consequence often referred to as 'silks'. The miller would darn his cloths as the fisherman mended his nets, and his daughter would dress her dolls in scraps of worn silk. In large mills bolters attained a considerable size, and by the late nineteenth century reels measuring 25 feet in length and 3 feet in diameter were not uncommon.

An alternative to the bolter was the wire machine or 'dresser', in which a series of rotating brushes swept the meal along a stationary cylinder lined with wire gauze. In both machines the flour was graded by the use of silk or gauze of varying degrees of fineness. The country miller was content to distinguish three or four grades, but in the large commercial mill, numerous intermediate grades of both flour and bran were recognised. In the North, where larger quantities of oats were grown, the mill often incorporated a kiln, the flue and cowl forming a distinctive feature of the building. After roasting, the oats were passed through a 'groat machine', in which the husks were removed by a fan. By the nineteenth century various other items of auxiliary machinery were in common use. 'Separators' and 'screeners' were employed to sift out dust and grit prior to grinding, and 'smutters' were used to clean the grain of a particularly prevalent type of mould. Rough dressing was sometimes performed by an assembly of oscillating sieves known as a 'jog-scry', and other tasks, such as cutting chaff and kibbling beans, were frequently carried out in the country mill.

During the nineteenth century mechanisation played an ever increasing part in the operation of the American mill, but the machines invented by Evans and his successors were principally concerned with the subsidiary processes. Although 'ending stones' were commonly employed for the preliminary loosening of the bran, the basic operation of grinding was still performed by the old process of 'low milling'.

It was realised by the merchant millers of the period that an enormous potential market existed for a pure white flour of standardised quality, if such a product could be offered to the public at a competitive price. As early as the sixteenth century, French millers had

An old mill at Orvenyes on Lake Balaton, Hungary, rebuilt in 1740 on a medieval site

experimented with the re-grinding of meal. This practice resulted in the production of a greater quantity of usable flour from a given sample of grain, and from this beginning the continental system of 'high milling' was developed. The meal was passed through several pairs of stones, set progressively closer together. Grading was effected at each stage in the process, and the finest flour produced was relatively free from bran, and unequalled for quality and whiteness. The poorer grades, on the other hand, were much inferior to the wholemeal flour produced in the British mill. 'High milling' reached its peak of development in Hungary during the nineteenth century, the mills of Buda Pest becoming world famous both for the quality of their finest products and for the staggering complexity of the processes by which these results were achieved. The system was prodigal of time and labour, and it was economically viable only in countries wnere the social structure provided markets for widely differing grades of flour. It was no doubt partly for this reason that the Hungarian system was never widely adopted in democratic America or in Britain.

From a purely commercial view point, 'old process' milling had many serious disadvantages. Not only did it involve the loss of a proportion of flour, discarded with the coarse bran at each grinding, but the colour of the final product was darkened by the presence of minute particles of bran which were extremely difficult to eliminate. It was essential that the flour produced should keep well, to survive the rigours of transport and storage, and this could only be achieved by the removal of the oily wheat germ. Millstones, which demanded skilled adjustment and constant maintenance, were proving less than ideal for the hard winter wheats which became increasingly popular in America from about 1850 onwards. It was realised that the adoption of some form of gradual reduction offered the only practical solution to these mounting problems, and the answer was eventually found in the 'roller mill'. This device employed cylinders of chilled iron, capable of very precise adjustment, in place of the traditional stones. The principle had been

A country watermill in Hungary

illustrated by Ramelli in the sixteenth century, but it was only after a concentrated period of trial and experiment in the early nineteenth century that a practical machine was developed in Switzerland. The 'roller mill', used in conjunction with elaborate separating and purifying machinery, made white flour of uniform quality available at last for mass consumption. It was not generally understood that the thorough elimination of wheat germ and bran entailed the loss of many of the vitamins present in stone-ground flour, and these deficiencies were later recognised, and to some extent made good, by the addition of vitamins and minerals to the finished product.

Although the first roller mills were served by water power, the new machines were equally suitable for operation by the steam engine. Steam had first been employed for milling on a commercial scale at the Albion Mills in London as early as 1784, and a steam mill was in operation in Pittsburgh, Pennsylvania, by 1808. But mill owners were reluctant to make the change where abundant water power was still available, and steam was not generally adopted until the latter part of the nineteenth century.

With the general acceptance of the new milling process, the production of flour by the traditional method rapidly declined. The small country mill was not equipped to meet the overwhelming public demand for white flour, and few small millers could afford to install the complex machinery required. In Britain, countless watermills were forced to give up the unequal struggle, and to end their days grinding feed for livestock. In the Southern States of America some millers were able to earn a livelihood grinding Indian corn or maize in place of the wheat of earlier days, but the era of the water powered corn mill had come to an end.

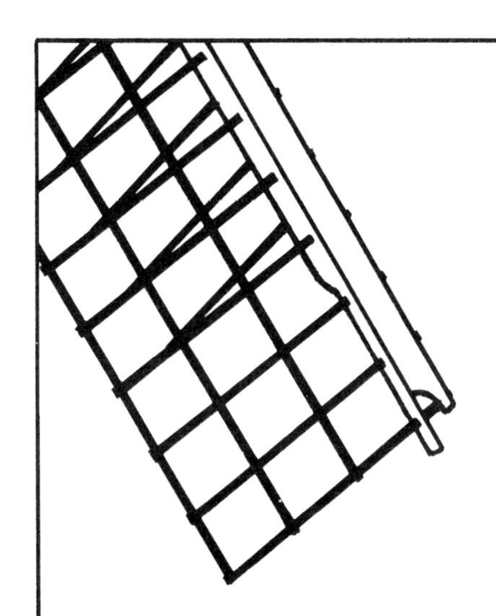

4 The Development of the Horizontal Mill

It is a curious fact that the Roman mill never completely supplanted its primitive forerunner. In the mountain fastnesses of the Alps and Pyrenees, and in scattered areas around the fringe of the Western World, the ancient horizontal Greek mill remained in use well into the twentieth century. Though established among isolated communities in the furthest corners of Europe, it was characterised by a remarkable uniformity of design and construction. No less strange, by comparison with the customs prevailing in feudal Europe, was the attitude adopted towards its use. This was no doubt due, in part, to the limitations of the machine itself, and in part to the traditional courtesy of mountain people inheriting few memories of Roman rule. Their mills were slow and inefficient, but they performed the task for which they were built, grinding for small communities which often consisted of no more than a single family group. There was no place here for Chaucer's grasping miller; indeed the farmer often worked the mill himself, filling the hopper with grain, opening the sluice, and leaving it to grind untended while he went about his work in the fields.

Often the door of the mill, like that of the church, stood open to all comers and by ancient custom no toll was demanded, the owner scorning to take a fee from his poorer neighbours. No villager was denied the use of the mill, though he was required, in courtesy, to ask the owner's permission. When mills were owned jointly, the partners would each be allotted a regular day for their grinding. No human institution is perfect, however, and arguments no doubt arose. Landowners conversant with lowland law attempted to enforce the alien rights of milling soke. But in many areas from the Faroe Islands to the Balkans, the old traditions of common use survived until the watermill finally passed out of service.

It has already been suggested that the Greek mill, despite its name, originated in the barbarian world, somewhere among the mountains to the north and east of India. This conjecture is based on the almost simultaneous appearance of the horizontal wheel in Asia Minor, and in China far away to the east. But wherever its place of origin, the Greek mill spread far and wide through the highlands of Europe and the Middle East. For motive power it required fast falling mountain streams, and its distribution may be followed, on a relief map, through

Norse mill at Skansen Open Air Museum, Stockholm, that formerly stood in the parish of Ås, Halland, near Gothenburg: note the wooden lever in the foreground, operating the sluice

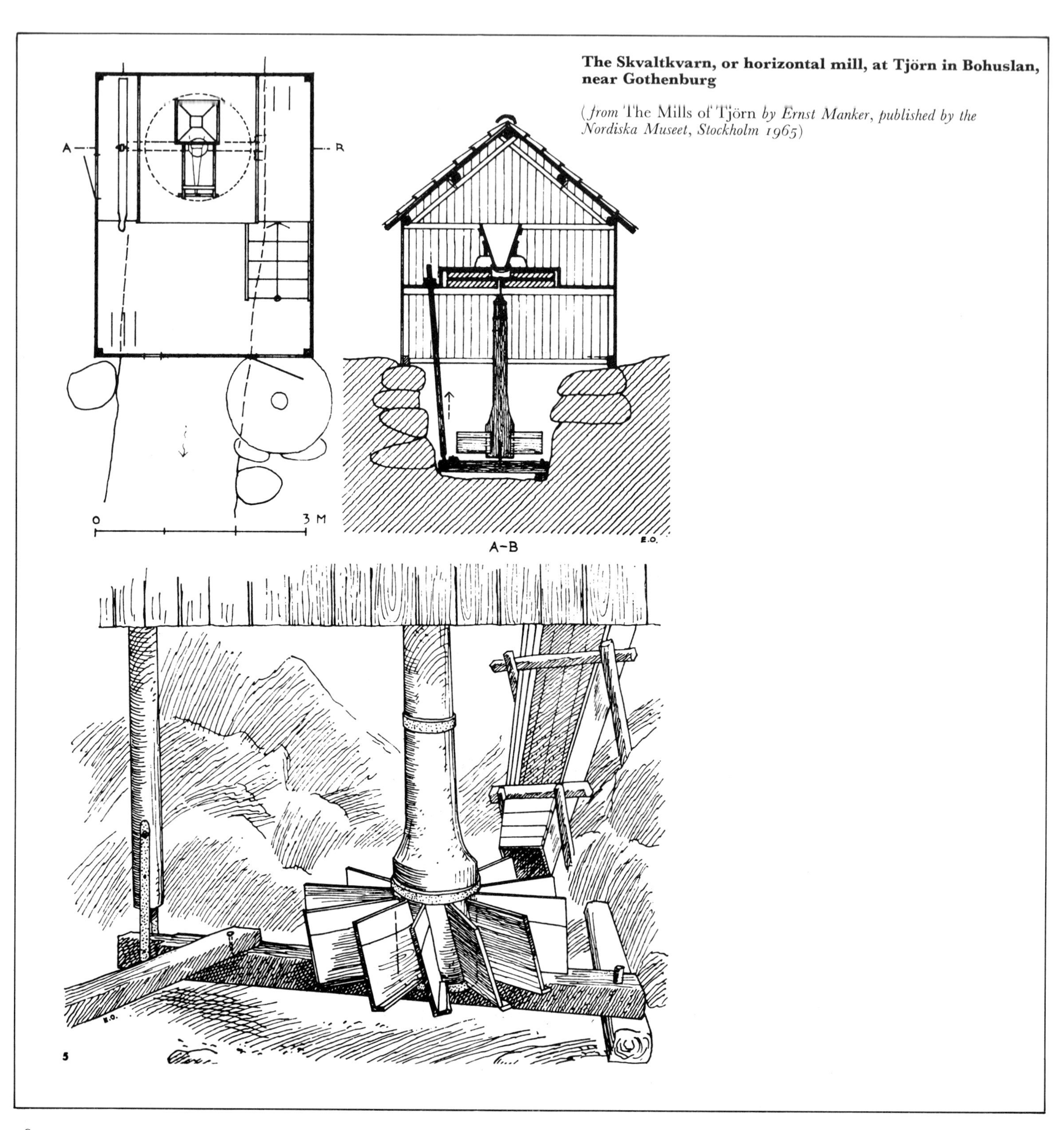

The Skvaltkvarn, or horizontal mill, at Tjörn in Bohuslan, near Gothenburg

(*from* The Mills of Tjörn *by Ernst Manker, published by the Nordiska Museet, Stockholm 1965*)

Asia Minor and the Balkans to Rumania and Switzerland, Southern France and Spain. Further south, it became established along the eastern shores of the Mediterranean, and in Cyprus and Crete. It probably reached the Baltic along the old overland trade routes from the Black Sea, becoming so firmly rooted in Scandinavia that in Northern Europe it came to be referred to as the 'Norse mill'. It was to remain the characteristic mill of Norway and Sweden, and Viking influence no doubt accounted for its introduction to Scotland and the Northern Islands. It was carried to Ireland at an early date, though whether through contact with the North, or along the ancient coastal route from the Mediterranean, remains undecided. A second wave of expansion was to follow centuries later, with the growth of the Colonial Empires. The humble Greek mill was relatively cheap and simple to construct, and was well suited to the needs of small pioneering communities. It was carried by European settlers to North America, where the slightly more sophisticated tub wheel, described and illustrated by Oliver Evans, enjoyed a considerable vogue. It was similarly introduced to South Africa, and on remote Boer farmsteads examples of the simple horizontal mill have remained in use down to the present day.

The typical Norse mill consisted of two elements: a basement or 'under-house', through which the mill stream flowed, and a 'meal-house' above. In its most primitive form the 'under-house' was enclosed by the rocky banks of the stream. In more developed examples walls were built up in rubble masonry to support the structure above, which took the form of a rectangular hut of stone or timber, with gable ends and a simple pitched roof.

A regional character was imparted by the use of local building materials. Scandinavian mills were commonly of timber, with walls formed from heavy planks set horizontally between squared corner posts. The under-house was built in dry-stone walling, and the steeply pitched roof was of thatch. Very similar mills were to be found in Southern Europe, some built of sawn timber and some employing the earlier log building techniques. With the development of more advanced methods of timber conversion, the use of weather-boarding and stud framing became common. In the Shetland Islands and in Northern Scotland, the mills, like the houses of the crofters, were of stone. The

Click Mill, Dounby, Orkney

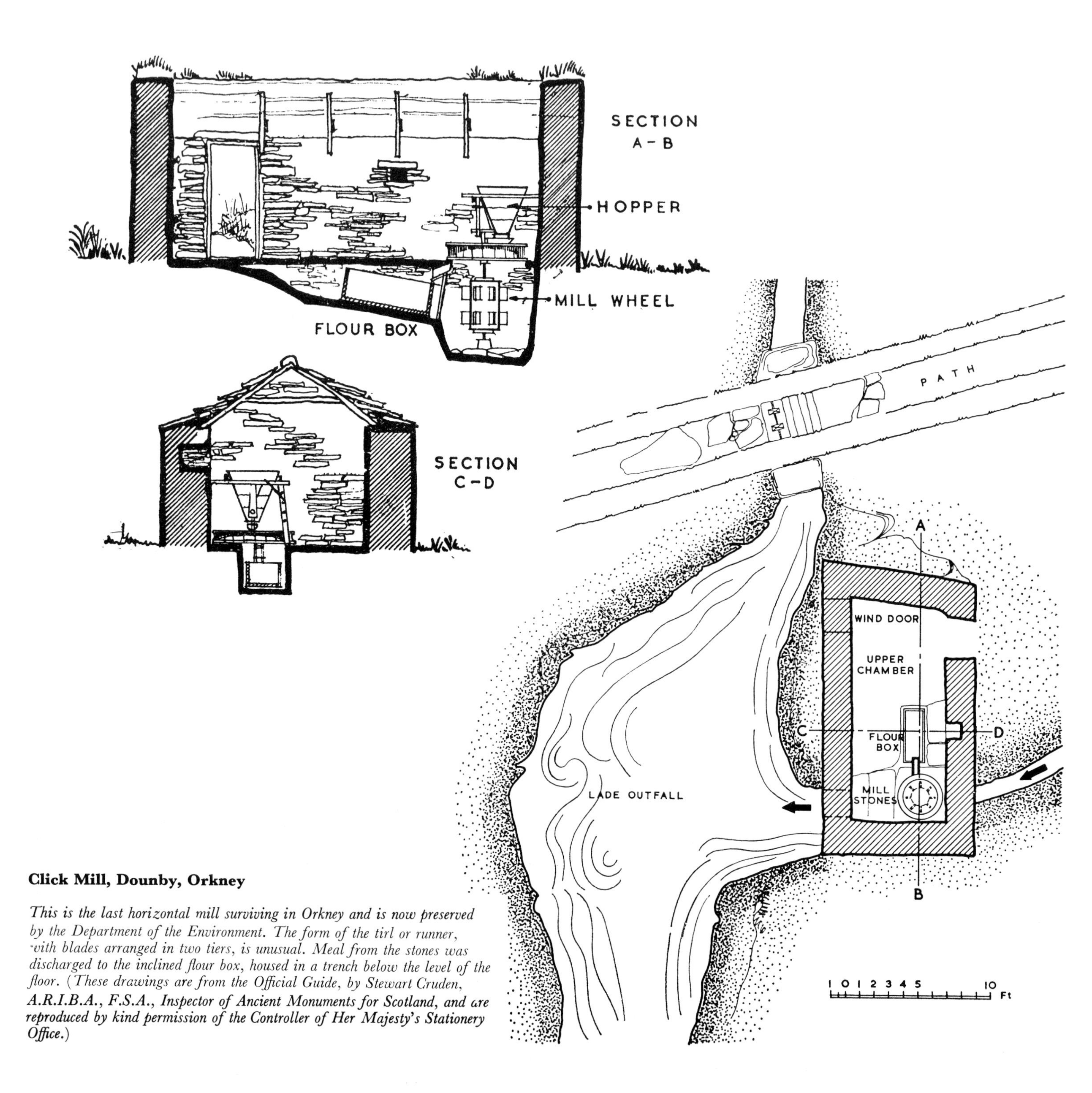

Click Mill, Dounby, Orkney

This is the last horizontal mill surviving in Orkney and is now preserved by the Department of the Environment. The form of the tirl or runner, with blades arranged in two tiers, is unusual. Meal from the stones was discharged to the inclined flour box, housed in a trench below the level of the floor. (These drawings are from the Official Guide, by Stewart Cruden, A.R.I.B.A., F.S.A., Inspector of Ancient Monuments for Scotland, and are reproduced by kind permission of the Controller of Her Majesty's Stationery Office.)

rubble masonry of the walls was laid dry, with a distinct batter towards the eaves, and there were no windows, light being admitted through the low doorway. Roofs were of turf or thatch, securely lashed down against the winter gales. These simple buildings were in perfect accord with the wild landscape in which they were set. In the Western Isles they were known as *Muilean Dubh* or 'black mills', after the traditional 'black houses' which they so much resembled. In the Faroes, old and new building techniques survived side by side. Early mills were of dry-stone walling expertly laid, while later examples were built of Norwegian timber, the roofs being traditionally covered with turf supported on an underlay of straw or birch bark. A single stream would often support a series of mills, set one below another on the mountainside. Where water was plentiful, no effort was made to conserve it, but in Scandinavia it was not uncommon for such a group of mills to be served by a single mill pond. In times of drought, the sluice would be opened at a pre-arranged time to allow all the mills to take advantage of the stored water.

Although no example of the Norse mill survives in Ireland, documentary evidence indicates that it was in general use there until comparatively recent times. Relics of early mills have come to light both by chance discovery and through careful excavation, fragments of timber recovered from the peat displaying a very high degree of craftsmanship, not equalled elsewhere in the north. Dams of timber point to the use of the mill pond, a refinement uncommon in Scotland and the Western Isles, and these factors suggest a separate derivation for the Irish mill. But the absence of a living tradition, and the strange circumstances in which some discoveries have been made, have combined to confuse the picture. The drainage of bogs and the scattering action of floods have also played their part, and fragments have been unearthed far from probable mill sites. Elsewhere later accretions have disguised the true significance of particular streams or pools. A holy well at Mashanaglass, long held in reverence by the local people, was found on excavation to conceal the remains of an early mill. Various saints of the Celtic church were traditionally credited with the foundation of water-mills, and this may account for the veneration of such a site. The horizontal mill was certainly in use in Ireland in the seventh century. According to legend it had been introduced three centuries earlier in the time of King Cormac macArt, who sent for a skilled man from 'over the wide sea' to build him a mill. His purpose was to free a favourite serving maid from the drudgery of grinding at the quern, and one is at once reminded of Antipater's famous epigram.

The Norse mill had no gearing, the basic mechanism consisting of a vertical wooden shaft, mortised at its lower end for a number of raking wooden paddles. These formed the horizontal wheel of the mill, known in the north as the 'tirl', the design of which varied considerably from district to district. In the Hebrides and Shetland Islands, flat wooden blades were commonly employed, set at a slight angle to the vertical. The number of blades varied between four and twelve and they were sometimes shrouded by a deep vertical rim of wood or sheet iron. The Moycraig tirl, excavated from a peat bog and now preserved in Belfast Museum, is of the more sophisticated design typical of ancient Irish mills, with paddles curved on plan and carefully shaped from the solid. Even greater refinement was achieved in the Alpine mill, where strongly cupped spoon-shaped paddles were used to form a 'tirl' resembling the Pelton wheel of later years. The vertical shaft terminated in an iron spindle, which passed through a bearing in the bed stone to support and drive the runner stone above. The weight of the runner stone was carried on a 'sile' or rynd of iron, shaped to engage with the squared

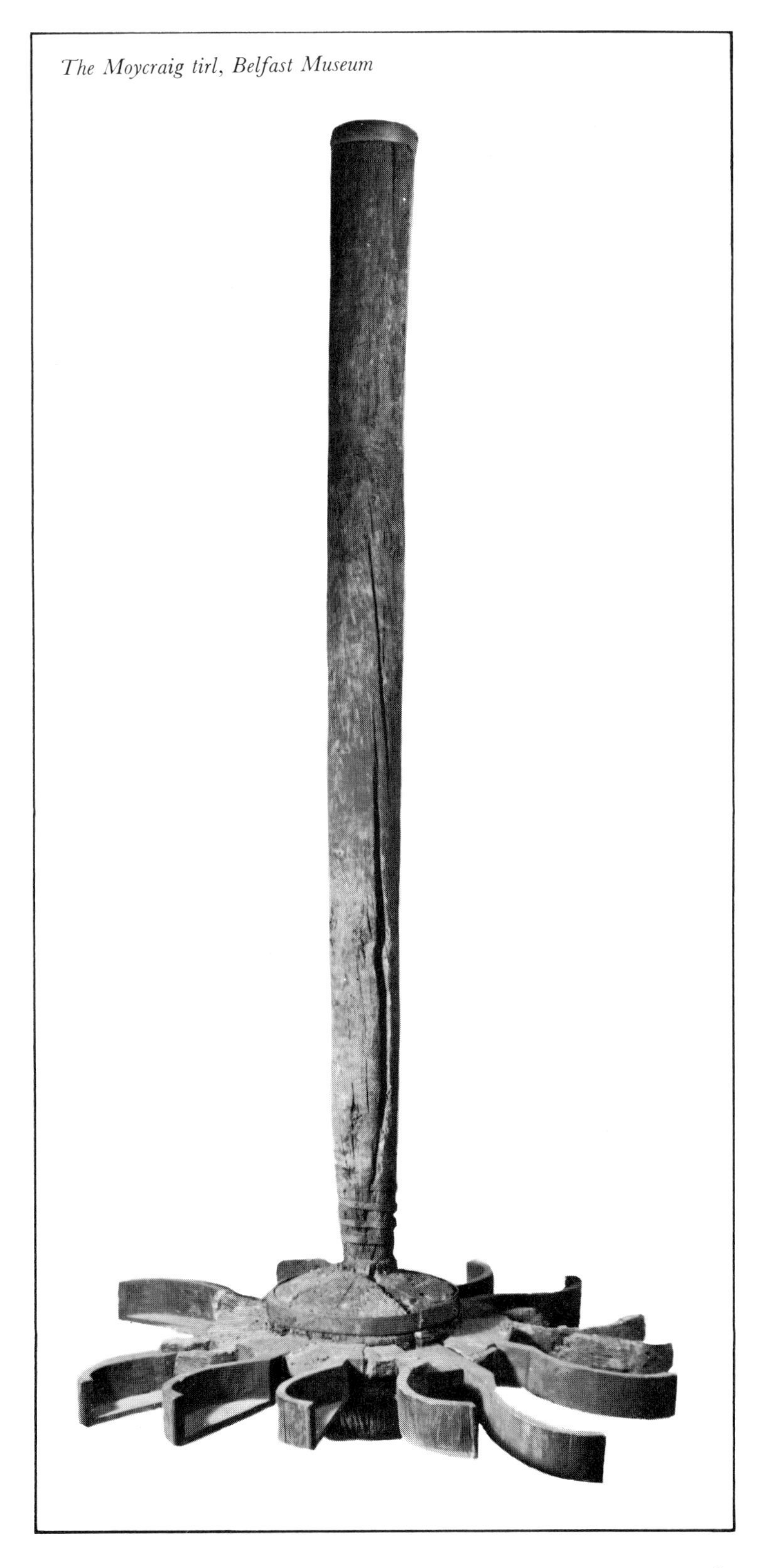

The Moycraig tirl, Belfast Museum

head of the spindle below. In the most primitive examples the shaft rotated on a bearing in the bed of the stream, either on a stone gudgeon or an iron spike set in the rock, and adjustment of the stones was effected with wooden wedges. Later millwrights adapted the method of tentering used in the vertical mill, mounting the thrust bearing on a movable timber beam, or 'sole tree'. One end of this could be raised or lowered by means of a vertical rod known as the 'lightening tree', which projected through the floor of the mill, and was held in position by a wooden key, or 'sword'. In the Faroe and Shetland Islands, where large timber was scarce, driftwood was always in demand, and shafts were sometimes cut from the spars of wrecked ships. Alternatively, the whole shaft might be of iron except for the wooden hub of the tirl.

A square wooden hopper hung from the rafters of the mill, feeding grain down to the stones through a tapering shoe. In the Faroe Islands, this was known as the *svini* or pig's snout. The grain was kept in motion either by a primitive clapper of stone or by the vibration of a flexible stick, loosely fitted in guides attached to the hopper and shoe. The

Aruba penstock—an aqueduct serving the mill at Lapithos, Cyprus

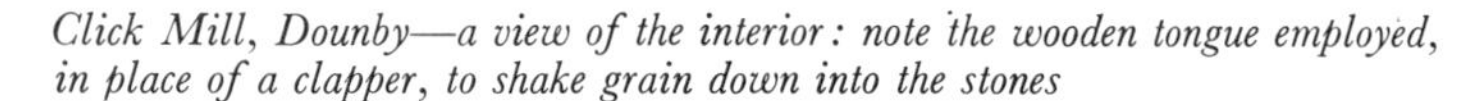

Click Mill, Dounby—a view of the interior: note the wooden tongue employed, in place of a clapper, to shake grain down into the stones

stick was weighted down with stones, so that it pressed against the runner stone, jumping along the uneven surface and shaking grain down from the shoe above.

The millstones were small and light, seldom exceeding three feet in diameter, and sometimes scarcely larger than those of hand querns. Without benefit of gearing, the runner stone revolved at less than half the speed of its counterpart in the more powerful vertical mill. Elaborate dressing was not employed, and grinding was slow.

The mill stream was channelled into an inclined trough or 'lade', arranged to discharge a jet of water against the blades of the tirl. From this narrow open trough, constructed either in timber or stone, the horizontal mill acquired its characteristic Scandinavian title of *skaltkvarn* or 'spout mill'. So long as the lade remained open, the velocity of the water could not be increased beyond a certain limit. This problem was overcome, in developed examples, by enclosing the lower end of the trough, thus producing a pipe in which pressure could be built up by a positive head of water. Lades of this type, covered by slabs of stone, appear to have been used in Ireland at an early date, the efficiency of the mill being further increased by the use of a restricted orifice or nozzle. More recent Scandinavian mills employed the same principle, the water being discharged through a tapered wooden chute, the lower end of which was enclosed by a boarded cover.

In the Near East, where water was commonly in short supply, the principle of the enclosed lade was carried to its logical conclusion with the development of the 'Aruba penstock'. This notable innovation consisted of a vertical shaft of masonry, built to contain a column of stored water, and fed from above by a leat. The millwright was thus able to take advantage of a considerable head of water, which was released under pressure from the base of the shaft. 'Aruba penstocks' were constructed in Palestine, Cyprus and Crete. Though few examples remain in use, these tall round towers, linked to the hillside by lofty aqueducts of stone, still present a most impressive appearance, dominating the mills themselves which huddle at their feet.

In remote areas, the design of the horizontal mill changed little during centuries of use. But in Southern France a series of developments took place which were to have the most far-reaching results.

The aqueduct serving the mill at Karovos, the last mill in Cyprus to work by water power

The mill at Karovos—a view from the valley

The mill at Karovos —two pairs of stones are driven by belting from the vertical spindle (right foreground)

The seventeenth century saw great advances made in the science of hydromechanics, and French scholars were particularly active in this field, which had received little attention since Classical times. As early as the fifteenth century, the horizontal mills of Toulouse, on the Garonne, were exciting the comment of travellers, and mills of this type were illustrated, late in the following century, by both Besson and Ramelli. One of these examples was driven by a drum-shaped rotor carrying helical blades, the other by a tirl equipped with spoon-shaped paddles remarkably similar to those of the Moycraig mill. A type of multiple mill developed in Languedoc aroused considerable interest during the middle years of the eighteenth century. The substructure was built in the form of a barrage across the stream and incorporated a series of tapered stone-built channels, each discharging to its own circular wheel pit. During the early nineteenth century, many horizontal mills were at work in France. Although commonly driven by tirls of conventional design, fitted with curved scoop-shaped blades, experiments were also made with various revolutionary types of rotor. One of these took the form of an inverted cone, equipped with spiral blades, and contained in a tapering chamber of masonry. The country miller, however, was content with a much less elaborate machine. Wherever an unrestricted water supply was available, with an adequate head, the simple horizontal mill had much to commend it. With few moving parts it was easy to build and maintain, and these advantages outweighed its relative inefficiency. One important innovation, foreshadowed by the introduction of the circular wheel pit in continental mills, was made by enclosing the tirl within a fixed circular shroud of timber. Tub wheels of this type were very widely employed in North America, and isolated examples remained in service well into the twentieth century.

The word 'turbine' had long been used in France to describe the horizontal mill, but was to take on a new meaning during the early years of the nineteenth century. A society of French industrialists had offered a prize with the object of improving the efficiency of the traditional machine, and in 1827 a contribution of major importance was made by a young engineer named Benoit Fourneyron. By admitting the water through a ring of fixed guide blades, curved in the opposite direction to the vanes of the runner, he reversed the direction of rotation of the water within the motor, thus producing a true turbine. In point of fact, Fourneyron's turbine was not the first to be used for milling. In 1743, a Doctor Barker had published designs for a machine of a very different type, a simple reaction turbine employing the principle of the revolving distributor installed over modern filter beds, the water passing down a hollow central shaft, to discharge in jets from tubes projecting on either side of the base. Water motors of this type were built, but for practical reasons their use was limited. Fourneyron's turbine, however, was an immediate success, and further development followed swiftly.

Two main classes of turbine were recognised, the reaction type in which the runner was completely filled with water, and the impulse type which functioned with the wheel passages partially filled. Each class was further subdivided to indicate the direction of the flow of water through the machine. Fourneyron's prototype was thus an outward flow reaction turbine and so successful was his original design that it was still in production, with little basic modification, at the close of the century.

Throughout the nineteenth century many engineers in both Europe and America were applying themselves to the development of the turbine. In 1826, Poncelet had propounded a theory for an inward flow machine, and during the late 1830's, motors of this type were installed by Howd in several New England mills. J. B. Francis carried out much

A horizontal mill from Le Diverse et Artificiose Machine *by Agostino Ramelli, 1588*

A horizontal mill fitted with two pairs of stones, driven through spur gearing. c. 1590, Veranzio

A Provençal mill from Architecture Hydraulique *by Bernard Forrest Belidor, 1737-1753*

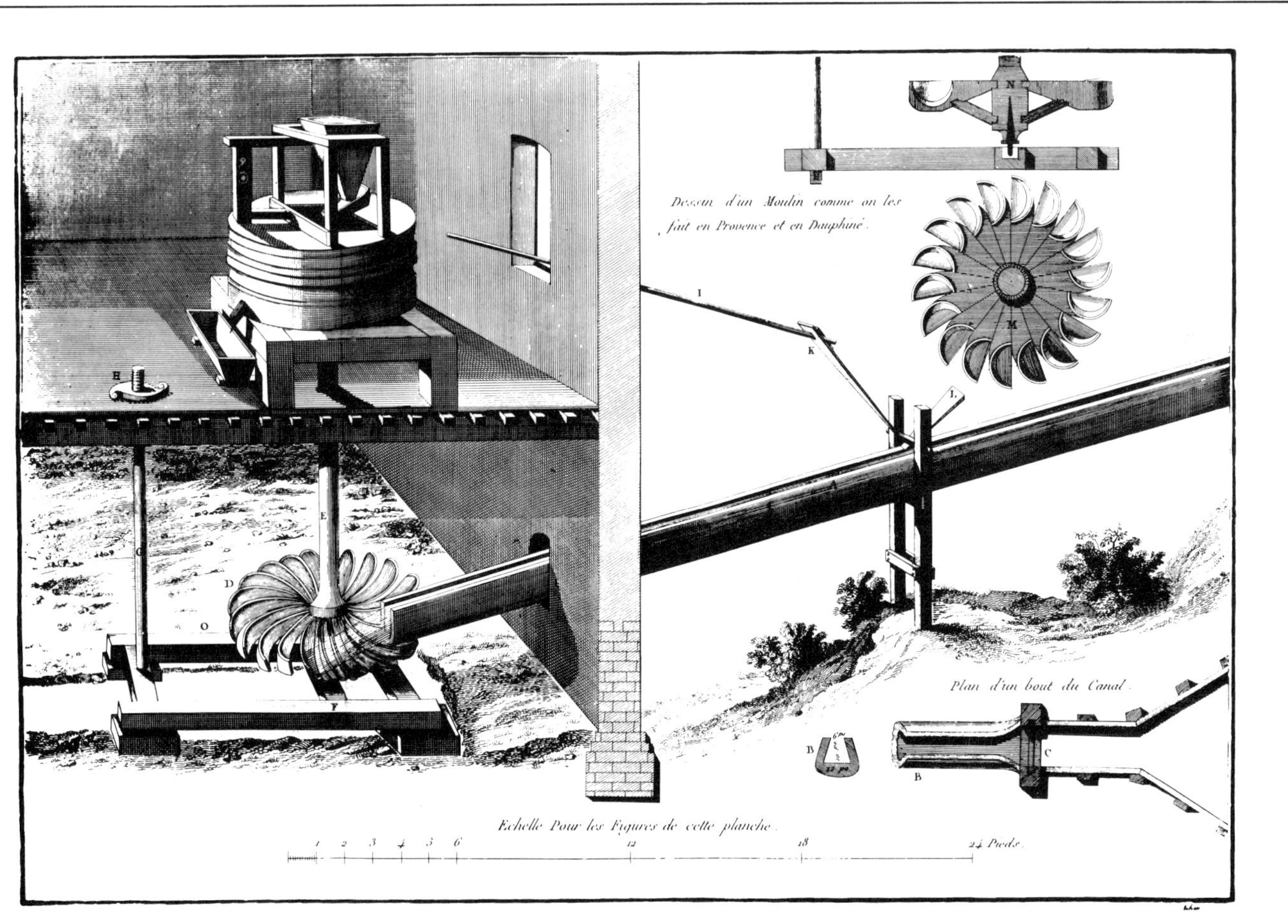

An American tub wheel, c. 1820, photographed at Bow in 1890

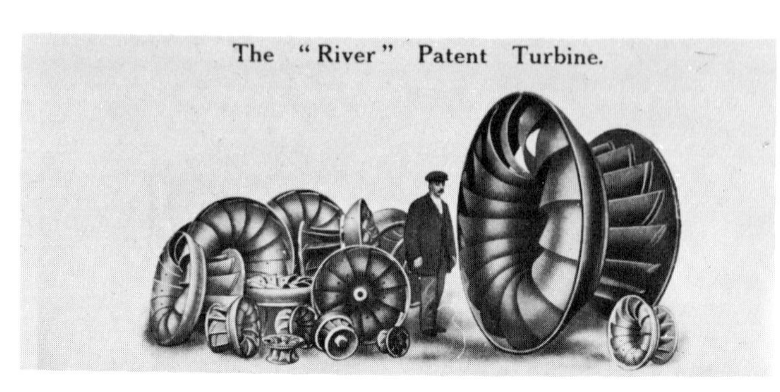

Early twentieth century turbine runners, from Messrs. Armfield's catalogue

A late nineteenth century American mill, equipped with rolls and powered by a turbine

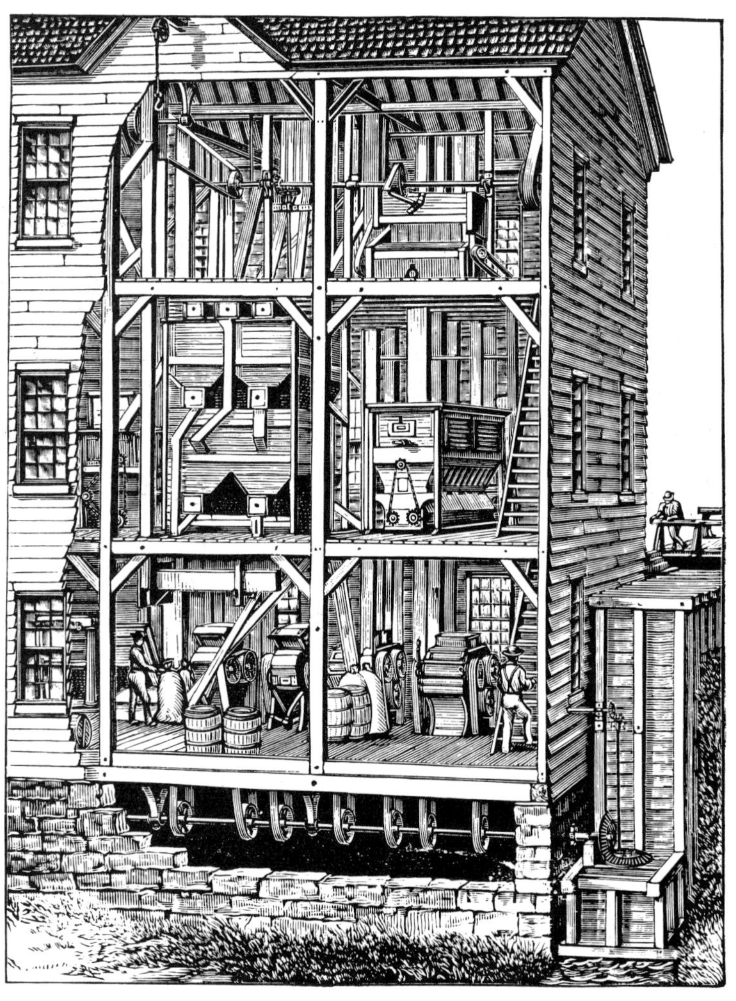

exploratory work in America, and Jonval's axial flow turbine had appeared in France by 1843. Girard designed the impulse turbine which bore his name, and Callon, Kaplan and many others contributed to the general advance, which opened up possibilities for the exploitation of natural power on a scale previously undreamed of.

Problems were quickly encountered in the regulation of speed, and it was found that efficiency declined rapidly under conditions of fluctuating water supply. These difficulties were overcome in various ways, perhaps most successfully by the use of adjustable guide blades in Thompson's inward flow reaction turbine.

As these new machines became commercially available they were at once put to work in mills and factories, where they were often able to show a considerable increase in efficiency over conventional water wheels. When wheels became in need of repair, they were frequently replaced by turbines, and while one may regret the destruction of the old machinery, many corn mills were undoubtedly given a new lease of life in this way. In the traditional watermill, the insertion of a turbine entailed a good deal of alteration to the mill race. Once installed, however, little maintenance was necessary provided that a suitable grating was fitted across the head race to prevent floating weed and debris from choking the runner. The Armfield Company of Ringwood in the New Forest, one of many firms specialising in work of this sort, equipped numerous English mills with their 'River' patent turbine. This was of the mixed flow reaction type, a modified version of the 'Francis' machine. Fitted with adjustable guide blades, it was notable for its handsome bell-shaped runner, the vanes of which displayed a striking double curvature. Power was commonly transmitted to the machinery of the mill by a lay shaft, driven through a bevel gear wheel mounted on the top of the vertical spindle. The roller mill, referred to in an earlier chapter, might employ belts and pulleys, but it was not unusual for conventional stones to be retained, mounted on a hurst frame of iron, and driven by direct gearing. Such installations frequently replaced the traditional arrangement of water wheel and stones, and in this way the turbine was adapted to 'old process' milling.

The production of stone ground flour was, however, to be one of the minor functions of the turbine, which was soon found to be ideally suited for the generation of electricity. Before the end of the century, various major hydro-electric installations, including the vast Niagara project, were in operation, and it is interesting to reflect on the train of events by which the humble horizontal mill finally ousted its Roman rival.

LEFT
Armfield's patent 'River' turbine, showing the runner (left) and adjustable guide blades (right)

BELOW LEFT
A mixed flow turbine runner, removed from a Hampshire mill

BELOW
Durngate Mill, Winchester—the iron hurst frame, installed c. 1917, with millstones removed. Power was transmitted through bevel gearing to two pairs of stones: note the handwheels for tentering the stones

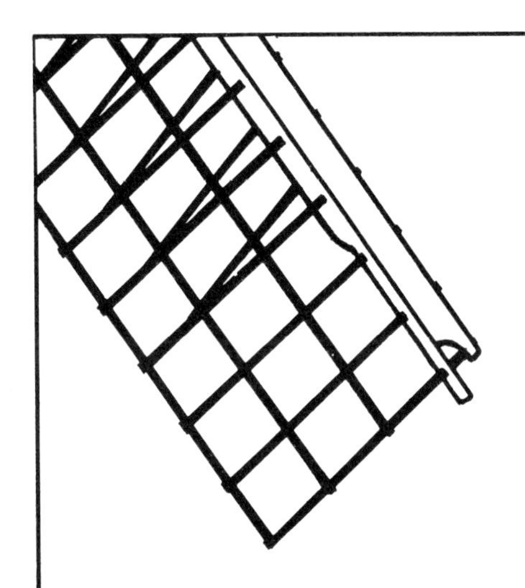

5 The Wind Powered Corn Mill

The origin of the windmill, like that of the watermill, is obscure. No one knows who first conceived the fantastic notion of harnessing the wind to grind corn, but primitive horizontal mills were certainly at work in Persia during the tenth century. They may, indeed, have made their appearance some 300 years earlier, but the evidence is inconclusive. The Persian mill was mechanically related to the horizontal watermill, the sails revolving on a vertical axle mounted in a square tower. The wind was directed on to the driving side of the sail assembly through diagonally opposed slots in the enclosing walls. According to tradition the principle was carried to the East by prisoners of Genghis Khan, and horizontal mills rigged with sails of matting became common in China, where they were used principally for irrigation. The Chinese millwrights dispensed with the enclosing tower, and achieved continuous rotation by the use of feathering sails. Designs for horizontal mills were published by the Renaissance engineers, but they never achieved popularity in Europe. A number were built, however, the best known being Captain Hooper's horizontal air mills, erected during the eighteenth century in Margate and London. The principle survives, in the West, in the familiar 'S' rotor employed in carriage ventilators.

The typical European windmill was almost certainly an independent invention of the Gothic North. In the twelth century this curious machine, perched on its single leg like some monstrous sciapod from the pages of a bestiary, must surely have seemed a work of genius bordering on madness. Even today the essential strangeness of the windmill exerts a powerful fascination, yet no wonder is expressed in the earliest certain documentary reference. This dates from 1185 AD and records without comment the existence of a windmill in the little village of Weedley in Yorkshire, let at a rent of 8s per annum. While it is likely that the principle was introduced from the continent, there is as yet no direct evidence for this, and it is at least possible that the sails of the first Northern windmills turned against the grey skies of England.

By the end of the century, the windmill had spread widely through Northern Europe and was common enough to warrant a Papal ruling on tithes. In the South it gained ground more slowly; it was firmly established in Italy by the early fourteenth century, but was still unknown

The Falcon *at Leiden, built in 1743*

RIGHT
A horizontal windmill mounted in a tower, c. 1595

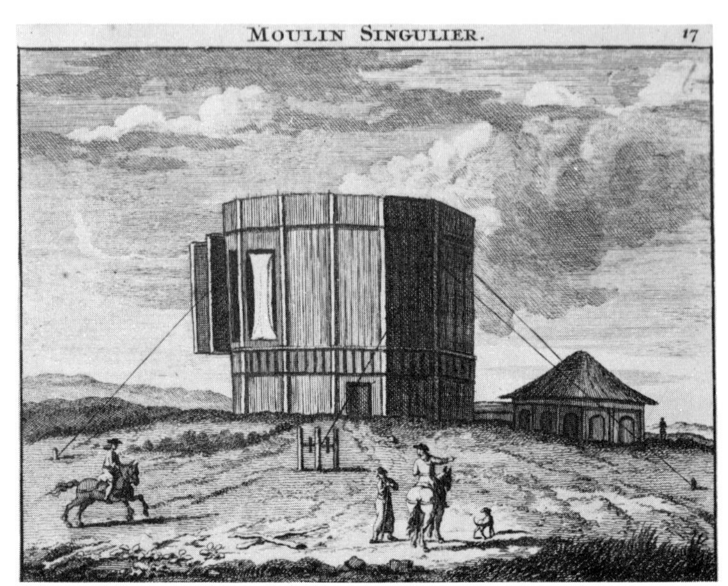

FAR RIGHT
A horizontal windmill, c. 1718

BELOW
A medieval post mill of the early fourteenth century, Decretals of Gregory, *British Museum*

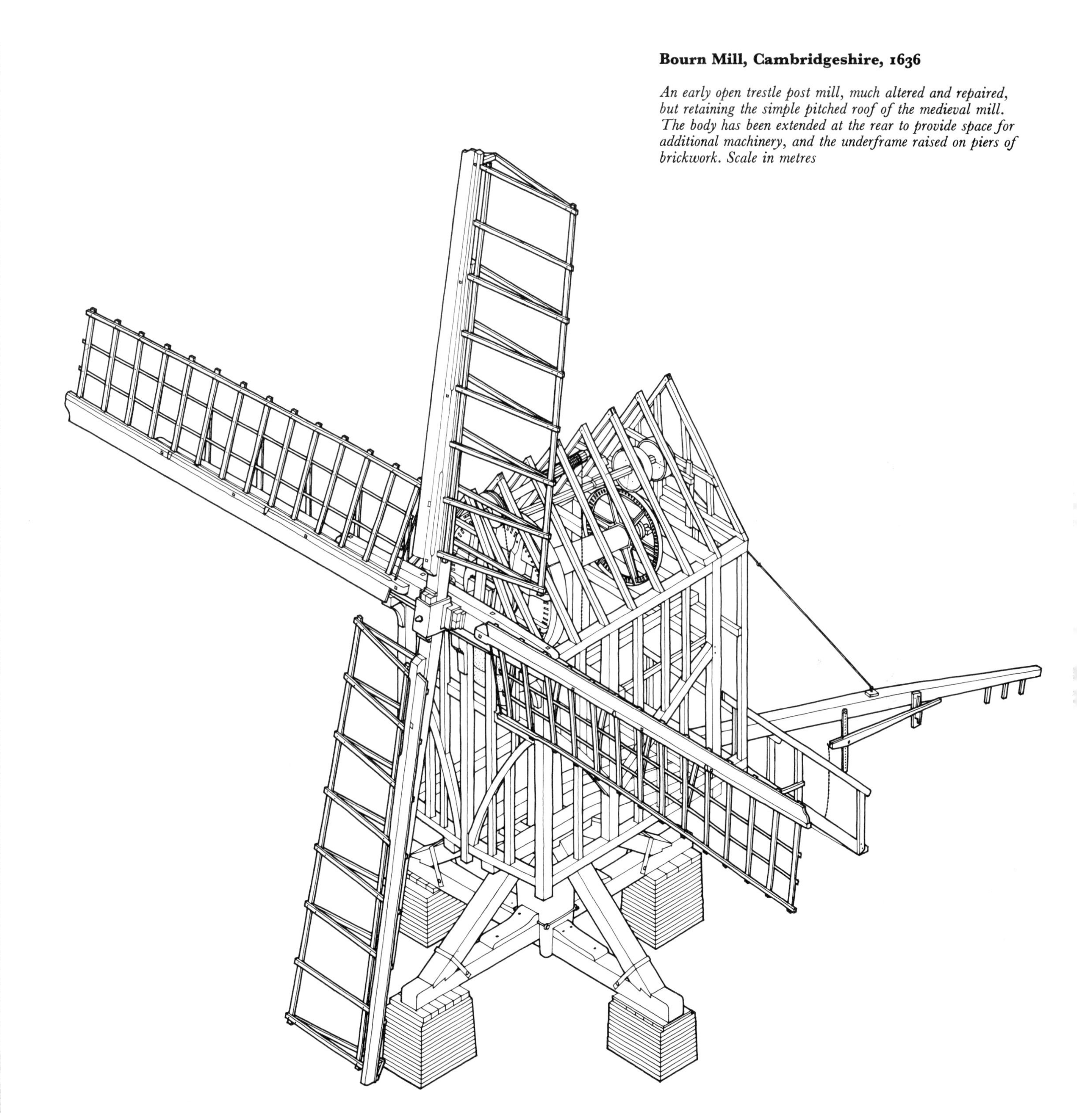

Bourn Mill, Cambridgeshire, 1636

An early open trestle post mill, much altered and repaired, but retaining the simple pitched roof of the medieval mill. The body has been extended at the rear to provide space for additional machinery, and the underframe raised on piers of brickwork. Scale in metres

LEFT
A post mill carved on a misericord in Bristol Cathedral, sixteenth century

in Spain 200 years later. German crusaders introduced the European mill to Asia Minor, and it is probable that it made its first appearance in the islands of the Mediterranean at about the same time. This represents a significant reversal of the prevailing flow of ideas from East to West, and lends strength to the theory that the European mill was developed independently of the very dissimilar windmills of Persia.

In medieval manuscripts dating from the thirteenth century onwards, drawings of windmills occur frequently, and it is evident that these early structures were 'post mills', differing little in basic design from examples built several centuries later. The 'post mill' was one of the most daring and ingenious works of the medieval carpenter. It was, in fact, a small timber framed building designed to balance and pivot on a single vertical support. This massive central post was set four square on a pair of oaken cross-trees, and strutted by raking 'quarter-bars' to provide a rigid cruciform underframe. The cross-trees were at first laid on the ground, or even buried for greater stability, but in later days it became customary to raise them on piers of brickwork or masonry, where they were less subject to decay. Traditionally the first posts were hewn from the trunks of living trees, though this has yet to be substantiated by archaeological evidence. It is interesting to note that the same origin was claimed for the posts of the spere screens in certain medieval houses. Some early drawings show mills apparently supported on unbraced posts, but it is probable that these are simply 'sunk post' mills, with the underframe buried below ground level.

The weight of the 'buck' or body of the mill was borne by the 'crown tree', a great transverse beam of oak, which rested across the top of the main post. The 'crown tree' was free to rotate in a horizontal plane, pivoting on a stout oaken 'pintle' turned from the solid timber. Later an iron gudgeon was fitted, or alternatively a complete cap and bearing of iron, known as a 'sampson head'. The ends of the 'crown tree' carried the frame of the mill, which was carefully put together with mortice and tenon joints, dowelled with hardwood trenails. Raking timber braces gave rigidity to the frame, and little ironwork was employed. Many old mills have subsequently been strengthened by the addition of iron tie rods, set up with bottle screws, to brace the framing of the buck. But such aids were unknown to the medieval millwright, who

BELOW
A post mill carved on a bench end at Bishops Lydeard, Somerset, sixteenth century: note the symmetrical sails of traditional form

LEFT
The junction between main post and cross-trees—Outwood Mill, Surrey

BELOW LEFT
Bourn Mill, Cambridgeshire, built prior to 1636

BELOW
Bourn Mill, the underframe

relied on traditional carpentry techniques, backed by the most sparing use of wrought iron fastenings. It was necessary to reduce the dead weight of the mill to a minimum, but at the same time to provide a structure with enough strength and rigidity to withstand both the vibration of daily grinding and the buffeting of winter gales. The mill had to be plumbed and balanced with great accuracy, so that the bed stone lay true and level no matter which way the buck was turned. Yet despite this most exacting requirement many ancient post mills remained in service for centuries, bearing witness to the extraordinary skill and craftsmanship of the men who built them.

Access to the mill was gained by a wide ladder, hinged from a platform at the back of the buck. Through the ladder projected a heavy timber beam, or tail pole, with which the whole structure could be turned. The miller would first raise the foot of the ladder with a lever and chain, the 'talthur', then set his shoulders to the tail pole and slowly walk the mill round into the wind. The ladder would then be lowered again, and the tail pole anchored to posts set in the ground around the mill.

The early post mill was capped by a simple pitched roof, like that which survives at Bourn, near Cambridge. Later it was found that curved rafters allowed more room for the brake wheel, and this led to the general adoption of the familiar rounded shape. An elegant variation was the ogee roof, in which the tops of the rafters followed a reverse curve to a pointed ridge. Most surviving post mills are clad in weather-boarding, a material possessing the qualities of strength and lightness, and are either tarred or painted white. Contemporary illustrations of medieval windmills show both vertical and horizontal boarding. The

BELOW
Ladder and tail pole at Chillenden Mill, Kent. The wheel in the foregound helps support the weight of the tail pole

BELOW RIGHT
Main post and crown-tree at Stevington Mill, Bedfordshire

Bourn Mill

Saxtead Green Mill, Suffolk

Tarred weatherboarding at Outwood Mill, Surrey, built in 1665 and still at work. According to tradition, villagers watched the Great Fire of London from the top of the mill

BELOW
Chillenden Mill, Kent—a late example of an open trestle post mill

former may still be seen in the tall post mills of Northern France, but in Britain horizontal ship-lap boarding was generally preferred for both walls and roofs. To exclude rain at the corners of the building, the mill-wright extended the boarding of the breast several inches past the corner posts, while the roof and sides were in turn extended to oversail the boarding of the rear gable. As long as the mill stood head to wind, this arrangement effectively weathered the vertical joints. Another development, not to be seen in early illustrations but common in surviving mills, was a shield-like extension of the breast, which gave a measure of protection to the main post and quarterbars below. In the Balkans this process was carried to its logical conclusion, deep skirts of boarding completely enclosing the underframe. The boarding of Dutch post mills was frequently fixed diagonally, a further refinement which tended to shed rain water at the corners of the buck. Both thatched and tiled roofs can still be seen, but because of their weight these materials were not favoured. When thin sheet iron became commercially available in the nineteenth century it was sometimes used for cladding, as at Cross-in-Hand in Sussex. In later years repairs to many mills have been less happily carried out in corrugated iron.

From the eighteenth century onwards the post mill was commonly provided with a round house, a circular enclosure built around the underframe, with a conical roof carried up to the underside of the buck. The round house increased the effective height of the mill, since it was necessary to allow working headroom below the cross-trees. It provided the miller with a useful store, and also protected the underframe from the weather. The walls were usually of brick or stone, according to the custom of the district, but some round houses were built of timber and raised on staddle stones to keep out the rats. Many old mills were modified in this way after long years of service, a fact attested by the presence of weather-beaten timbers within the roof, indicating that post- and cross-trees were once exposed. There were two doorways, on opposite sides of the round house, and if the turning sweeps made one impassable, the miller used the other. The main post passed upwards through a circular opening at the apex of the roof, which was sometimes tiled, but more often boarded and covered with tarred canvas.

Cross-in-Hand Mill, Sussex

Holton Mill, Suffolk

The medieval post mill had been designed to accommodate a single pair of stones. When it became necessary to house additional machinery, the buck was often extended to enclose the platform at the top of the ladder. Alternatively, space was provided by building out compartments from the sides of the mill, and this was the method usually adopted in Northern France. Machinery could not be installed in the round house because of the difficulty of transmitting a drive from the revolving buck above. In a modified version of the post mill, which became firmly established in Holland, this obstacle was overcome by the use of a hollow post, which allowed a vertical drive shaft to pass down into the base of the mill. These 'wip mills' were primarily associated with drainage, and are described in another chapter, but some were built as corn mills. In Britain a solitary example still stands on Wimbledon Common.

No early medieval post mills survive, and this is not surprising when consideration is given to their timber construction, and to the stresses and dangers to which they were constantly exposed. The oldest examples still standing were certainly in existence early in the seventeenth century, when the first dated inscriptions occur, though they may well be older. By this time, however, a new type of mill had come into being, and the ruins of late medieval tower mills have outlasted their timber counterparts.

The tower mill was built of masonry or brickwork, surmounted by a timber roof. This cap, which contained the 'windshaft' or axle on which the sails revolved, alone turned to face the wind, and the drive was transmitted to the millstones below by a central vertical spindle. The tower mill appears to have been a European invention of the early fifteenth century, and from that date onwards it provided an alternative to the timber post mill, though it never supplanted the earlier machine. The oldest British example has been identified at Burton Dasset in Warwickshire. Its original use forgotten, this round stone tower had been known to generations of villagers as 'The Beacon'.

Outwood Mill—interior of the roundhouse, looking up at the floor of the buck above. Grease glistens on the steady bearing, where the main post passes upwards between the shears

A village corn mill near Leiden, an example of the Dutch hollow post mill

FAR LEFT
French hollow post mill

LEFT
The date '1627' carved on a timber at Pitstone Mill, Buckinghamshire—the oldest surviving inscription in an English mill

BELOW
Tower mills and post mills of the late sixteenth century, Strada

RIGHT
A tower mill at La Mancha, Spain (restored)

BELOW RIGHT
The mill at Tés, Hungary—a tower mill dating from 1840, but retaining the primitive form of earlier centuries

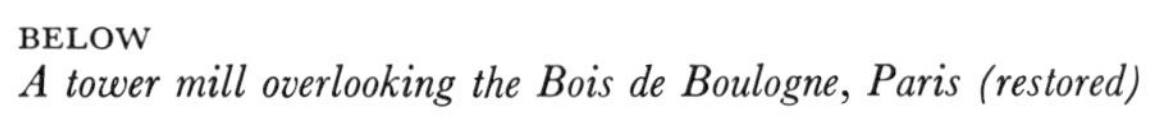

BELOW
A tower mill overlooking the Bois de Boulogne, Paris (restored)

The early towers were squat and cylindrical, like those still to be seen in Spain. They were capped by conical roofs of thatch or tiles, with small projecting dormers to weather the windshaft. In cities they were frequently set on the walls, above the crowded streets, and castle mills commonly occupied a similar position. It was necessary for both sails and tail pole to pass within reach of the miller standing on the ground below, and this practical consideration was to limit the size of the simple tower mill for centuries to come. The shape of the tower was dictated to some extent by the angle of the windshaft. In early mills this was mounted horizontally, and a cylindrical structure was necessary to enable the sails to pass clear of the wall face below. In some cases the upper half of the tower was even corbelled out providing greater clearance at the base, and a series of monumental stone towers of this type survives in Brittany. Later millwrights balanced the weight of the sails by inclining the windshaft, producing a result both visually and mechanically satisfactory, and making it possible for towers to assume the characteristic tapered form familiar today. Most were circular on plan, but a few were polygonal. They were built of local materials, and in the north they were often rendered or tarred against the driving rain. These truncated cones of masonry or brickwork were immensely strong and inherently stable; many towers stand intact long after sails and caps have rotted and fallen.

In flat country, and more particularly in towns, there was a demand for taller mills. Difficulties of operation were overcome by the provision of a stage, built out on brackets from the side of the tower. From this boldly projecting gallery, the miller could adjust the sails and heave round the tail pole. The finest mills of this type were built in Holland, where many can still be seen. They soar above the surrounding streets, heights of 100 feet to the cap being not uncommon. Few mills of this size were built in Britain until the invention of the patent sail and fantail had removed the primary need for the stage. No doubt this was partly due to the different physical character of the country, and partly to the fact that extra space was of more value to the Dutch miller, who continued to live in the mill long after his British counterpart had moved out to the mill house.

The tower mill at Batz, Brittany—a richly ornamented example of an ancient building type. The earliest surviving petit pied *mill dates from the mid fourteenth century. Sails constructed without hemlaths were not unusual on the continent*

LEFT
Bembridge Mill, Isle of Wight, early eighteenth century: the weather side of the stone tower is rendered against the south-westerly gales

BELOW
The tower mill at Kiskundorozsma, Hungary, built in 1821

Sash windows to the miller's quarters in the base of The Falcon

A tower mill with stage (Stellingmolen)—The Falcon *at Leiden*

RIGHT
Brick base to the mill at Herne, Kent

BELOW RIGHT
Smock mills clad in shingles at East Hampton, Long Island, U.S.A.—from Picturesque America, *by W. L. Bryant, 1872*

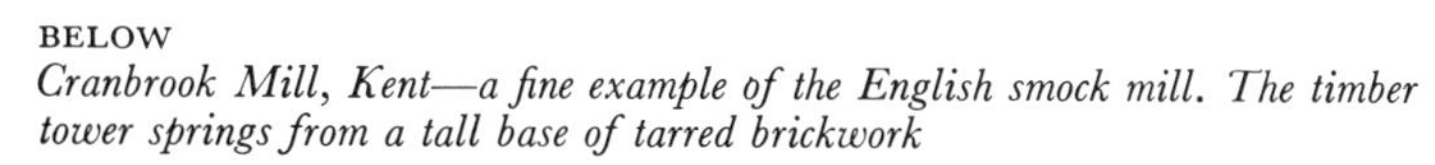

BELOW
Cranbrook Mill, Kent—a fine example of the English smock mill. The timber tower springs from a tall base of tarred brickwork

Herne Mill, Kent, 1789

This fine Kentish smock mill was raised on a two storey base of brickwork during the nineteenth century, and remains in service as a mill, though no longer powered by the wind. The structure has been repaired and strengthened over the years, and obvious additions, such as steel joists inserted to reinforce the cap frame, and timbers strengthening the cant posts, have not been included. The drawing shows only the main framing of the sails, or sweeps. For the sake of clarity the shutters and striking gear have been omitted, together with the intermediate stud framing to the front of the mill. Part of the missing guardrail round the stage has been restored. Scale in metres

The third basic type of windmill to be found in Northern Europe was the 'smock mill', which carried on the medieval tradition of timber spire building. The 'smock mill' took the form of a tapering tower, usually octagonal on plan. At each corner a great 'cant post' extended to the full height of the mill, its foot resting on a timber cill, its top mortised to a circular curb. The sides were subdivided by horizontal framing and strengthened by tiers of diagonal bracing, thus producing the stable and rigid structure necessary to support a revolving cap. In nineteenth century British 'smock mills', the wall panels were cross-braced, and similar elaborate cross-bracing characterised the great timber 'polder mills' of Holland. The floor joists were carried by pairs of massive binders, spanning from one side of the mill to the other, each pair arranged at right angles to those of the floor below. The cill was usually set on a plinth of brickwork, the original purpose of which was to form a solid and level foundation for the mill. In the nineteenth century, brick bases of one or two storeys became common, and the height of the mill was no longer restricted by the availability of large timber for 'cant posts'. A number of old mills were raised bodily on to new bases, a daunting operation but one to which the millwrights of the period were quite equal. Smock mills were clad in weather-boarding, and took their name from the bleached linen smock now no longer seen, but once the traditional costume of the British countryman. They were either painted white, as the name suggests, or else tarred, but in Holland they presented a very different appearance, both tower and cap being thatched. Another characteristic of the Dutch smock mill, particularly noticeable in the South, was the handsome waisted shape, achieved by the use of curved 'cant posts' rising from a base of brickwork.

Both tower and smock mills were surmounted by caps of timber, the shapes of which varied widely over the years. The cap was built on a stout horizontal frame, which carried the weight of the sail assembly and windshaft. It revolved on a circular curb of timber, which formed the top of the tower. Early curbs were 'dead', the millwright relying on the liberal application of grease to reduce friction under the turning cap. In later years it became customary for a 'live' curb to be fitted, the cap frame revolving on rollers of hardwood or iron. Centering was effected by a set of truck wheels, mounted in a horizontal plane and bearing against the inner face of the curb. The principal components of the cap frame consisted of two longitudinal members, the 'sheer trees', and three transverse beams. At the front of the cap was the 'weather' or 'breast' beam, which carried the neck journal of the windshaft; on the centre line was the 'sprattle' beam to which the top bearing of the main upright shaft was secured; towards the rear of the cap was the tail beam, on which the thrust bearing of the windshaft was mounted. Above this basic framework, caps developed strong regional characteristics, and in Britain several distinct types were in use during the eighteenth and nineteenth centuries. In Kent a boarded cap like the curved roof of a post mill was favoured, while a variant of the same type, common in East Anglia, had the sides strongly bowed, producing a shape reminiscent of an upturned boat. Also popular in the Eastern counties were domed caps of boarding covered with painted canvas, and, most handsome of all, ogee caps surmounted by ball or acorn finials.

Windmills possess a functional beauty independent of any applied decoration. In Britain, however, early examples frequently display some ornamental feature, either a strongly moulded bracket below the breast beam, or a graceful hood over the door. In some cases the workmanship is surprisingly fine, as at Pitstone mill in Buckinghamshire

ABOVE (AND COVER)
Pitstone Mill, Buckinghamshire—later rebuildings have changed the appearance of this ancient open trestle post mill

OPPOSITE
De Bleke Dood, *a smock mill at Zaandijk, Holland; the tower and cap are thatched above a base clad in weather-boarding*

BELOW
Pitstone Mill—the handsomely moulded main post

ANNO

A carved and gilded 'beard', or date board, preserved in The Falcon, *Leiden*

which bears the date 1627, but may well be older. Here the main post is beautifully turned and moulded, and the framing of the body displays elegant stop chamfers which would not have shamed the manor house or the village church. In general the British millwright was content with crisp paintwork and painstaking workmanship and he did not concern himself overmuch with decoration, though brightly coloured fantails were not unusual. In Holland, by contrast, it was customary for all exposed woodwork to be gaily painted in traditional colours. In wip mills, the predominant colour might be red or blue, green or yellow, the framing being picked out in white, and the 'poll end' embellished with a coloured star. In tower and smock mills, decoration was concentrated around the gable ends of the cap, providing focal points of colour against the sombre grey-brown background of the reed thatch. A carved date board, or 'beard', often elaborately decorated with scroll work, was fixed below the windshaft, the whole scheme recalling the 'gingerbread work' applied to the stern of a sailing ship. Again bright primary colours were used, and details were picked out in gold leaf if cost allowed. A very distinctive shade of deep green was particularly associated with the Zaan district, and green painted boarding was sometimes used as cladding in place of thatch.

The windmill, as perfected by the British millwrights of the nineteenth century, was an extraordinarily sensitive and sophisticated machine. The vagaries of the wind, always a fickle and uncertain source of power, made it at once more complicated and less reliable than the watermill, and demanded very different qualities in the miller. Like the master of a sailing ship he depended on the wind for his livelihood, but he knew that the power that drove his mill could easily destroy it. If he hung on to his canvas too long in a rising wind the mill might 'run away' beyond control of the brake, and burn to the ground before his eyes. If he could keep the stones supplied with grain throughout the storm, he might ride it out in the mill, but once the hoppers became empty a fire would almost certainly ensue. In such an emergency the miller would try to 'quarter' the mill, by putting the fantail out of gear, and working the cap round by hand until the sails were at right angles to the wind. In rough weather there was always the risk of a sudden squall 'tail winding' the mill and summer thunderstorms, with their dramatic shifts of wind, presented a peculiar hazard. Windmills were primarily designed to withstand a frontal attack, and in such adverse conditions it was not uncommon for the caps of tower mills to be dislodged, and for post mills to be thrown bodily over. By contrast, periods of calm brought enforced idleness and just as the seaman, when becalmed, would watch other ships on the horizon for a hint of wind, so the miller would watch the sails of a distant rival for signs of movement. When at last the wind blew again he would rise early and work late to make full use of it. If circumstances required, he would think little of working on through the night, and would even risk prosecution by starting the mill on a Sunday if providence sent a fair wind.

Windmills, like ships, were feminine, and in Holland many bore names, always preceded, as ships' names should be, by the definite article. Thus *The Falcon* still towers above the old city of Leiden, while elsewhere are to be found mills bearing such names as *The Star*, *The Trusty Watchman*, *The Hope* and *The Young Hero*. This practice does not seem to have been followed in Britain to any great extent, mills usually taking the name of the village they served. If there was a chance of confusion they would be distinguished by the name of the miller (past or present), or by some descriptive adjective. Thus *The White Mill*, and *The Black Mill* were probably the names most frequently encountered. In the nineteenth century, names such as *The Good Intent Mill* and *The Zenith Mill*, did appear; solid and respectable titles obviously chosen to inspire confidence. But whatever the name, the miller would always refer to the building familiarly as 'she'.

In the daily running of the mill, certain time-honoured customs were observed. Whether it stood on a lonely hill top or overshadowed the village street, the building was a prominent landmark, and when the

The cap of a Dutch tower mill, showing an ornamental dateboard in position below the breast beam

The foundation stone of The Falcon

The grim sign board of the curiously named De Bleke Dood *mill at Zaandijk*

sails were at rest their position conveyed useful information to the passer-by. When the day's work was over, or the mill was stopped for any considerable length of time, they were set diagonally, so that the strain on the stocks was equally shared. Before starting work in the morning the miller would 'strike up' the sails into the position of the true cross. In the days of the common cloth spread sail this was a necessary preliminary, but the action acquired a religious significance, and was continued by many millers long after the introduction of the patent sail, being known in Britain as 'miller's pride'. During the working day an upright cross might be a sign of mourning, and in some parts of the country a complicated code involving the removal of shutters was employed to indicate a death in the miller's family. In Holland certain subtle variations were observed. Celebration and mourning were distinguished by stopping the sails either just before, or just past, the upright position. Like ships, windmills were 'dressed overall' with bunting at times of rejoicing, and brightly coloured swags and garlands were hung between the sails, making a brave display. At such times the national flag was proudly flown from an ensign staff at the back of the cap, and in many old mills one may still find flags stowed away ready for future use.

In little more than a life's span, the windmill, like the sailing ship, has become a thing of the past. Each was an outstanding example of the use of natural power, and there were many points of resemblance between the two. An old post mill coming quietly to life in a steady breeze invites comparison with a sailing ship getting under way. The miller performs his preliminary ritual, pulling down the sails in turn and spreading the cloths, or adjusting the springs if the sails are shuttered. When all is ready the brake is pulled off and the sails begin to turn, slowly at first, then with steadily increasing speed. As they sweep past the windows, light and shadow alternate within the mill. The structure trembles slightly, like a ship feeling the first faint motion of the open sea, and the impression is strengthened by the presence of the great post, rising like a mast through the centre of the buck. On climbing the ladder to the floor above, one finds that the whole mill is alive with quiet movement. The interior is dominated by the bulk of the turning brake wheel, which fills the gable end, and by the massive central windshaft. The machinery is surprisingly silent in operation, but a number of small sounds combine to produce a low nurmur of activity. It is possible to pick out the rush of the circling sails, and the clucking of wooden gears as the stone nut spins against the rim of the brake wheel. The timbers of the buck creak as a gust strikes the mill; there is the dry rustle of grain as the miller fills the hopper over the stones, and loudest of all, the insistent rattle of the damsel against the shoe. A curious rising whisper comes from the stones themselves, where the runner sings like a ponderous top in its wooden casing, and a thin white smoke of flour drifts upwards to hang in the air above. The mill is full of the smell of fresh meal. White dust lies on every ledge, and hangs in the cobwebs under the rafters, while the worn floor boards and ladder rungs gleam as if wax polished through the action of meal trodden underfoot. Unhappily the smell and movement of the working mill cannot be recreated, no matter how carefully the fabric may be preserved, and the few old windmills which remain at work in England, grinding corn in the traditional manner, provide an irreplaceable link with the recent past, evoking most movingly the vanished rural culture of our fathers.

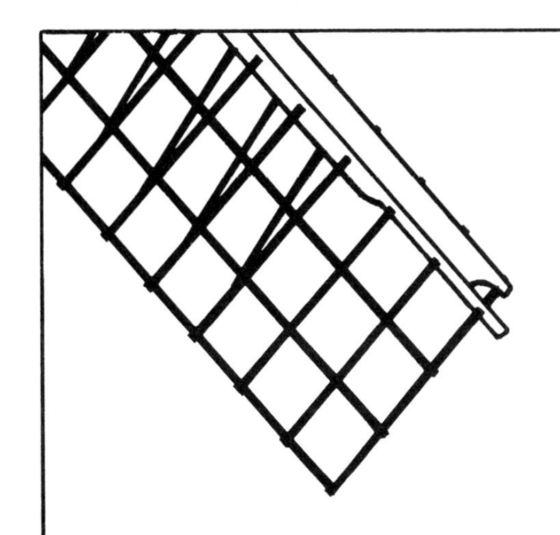

6 The Machinery of the Windmill

The sails of the windmill constituted a wheel of spectacular size, even by modern standards. Turning briskly in a fresh breeze, they formed a dramatic focus of interest in a static landscape, and the working mill dominated the countryside for miles around.

In Northern Europe sails were of canvas, spread on light wooden frames. They were carried on stocks, stout tapering spars crossed at right angles and mortised through the end of the windshaft. Lateral support was provided by 'sail bars', short transverse battens mortised through the stocks, like the rungs of a ladder, their ends linked by longitudinal 'hemlaths'. Medieval sails were of constant pitch, and presented an equal area to the wind on either side of the stock. The cloths were in two widths, threaded in and out between the 'sail bars' and not spread across the face of the sail, in the manner of later years. The symmetrical medieval sail was replaced in the North by the type known today as the 'common sail', which had most of its surface area arranged on the trailing side of the stock. A narrow windboard fixed along the leading edge directed the flow of air on to the cloth, the inner end of which was secured to a transverse iron rung. The canvas was spread down the length of the sail by means of 'pointing lines' attached to its edge. With these the miller was able to reef his sails to suit the strength of the wind, from full sail down through 'first reef', 'dagger point' and 'sword point', the latter leaving only a narrow triangle of canvas exposed. When the mill was at rest the cloths were furled and twisted around the framing of the sail.

As mills grew larger, and sails increased in size, it became customary to construct each sail on a lighter tapering spar or 'whip', strapped and bolted to the face of the stock. The mortises where the stocks passed through the windshaft had always presented points of weakness, allowing the penetration of rain water. With the introduction of cast iron this problem was at last overcome, the end of the shaft being replaced by a massive casting known as a poll end or 'canister', in which the stocks were secured by wedges. In the North East of England an alternative solution was adopted. The windshaft was terminated by a heavy iron cross, and to this the sail backs were bolted, the stocks being omitted altogether. In Holland, iron stocks were widely used, but the British millwright preferred the traditional timber construction for these components.

Timber windshaft with mortised sail stocks in a Dutch mill

The common sail was simple and efficient, and it remains in use on the continent to the present day. Its chief disadvantage lay in the fact that it could not be adjusted without first stopping the mill, and then bringing each sail down in turn to within the miller's reach. As the wind increased so did the speed of the mill, and in a sudden storm control could easily be lost. In 1772, Andrew Meickle designed a sail which was capable of 'spilling' the wind during squalls, and 'spring sails', based on his invention, were very widely adopted in Britain. The sail cloth was replaced by a series of rectangular shutters, hung on pivots and linked by a long connecting rod or shutter bar which ran the length of the sail. An adjustable spring maintained the shutters in the closed position against normal wind pressure, but allowed them to open during gusts. The actual mechanism differed widely from mill to mill, and both leaf and coil springs were employed. The shutters were of wood, or of painted canvas stretched on a light supporting frame. Spring sails were not without drawbacks; they were less powerful than common sails, and it was still necessary to stop the mill to adjust the tension of the springs. But they offered the miller a degree of automatic regulation unknown before, and it was not unusual to see a mill fitted, by way of compromise, with one pair of spring sails and one pair of the common type.

During the closing years of the eighteenth century, British millwrights strove to develop a method by which sails could be adjusted without stopping the mill. 'Roller reefing' gear achieved some popularity, and this mechanism paved the way for William Cubitt's patent sail, which appeared in 1807. Patent sails were fitted with conventional shutters and shutter bars, but these were controlled by an iron striking rod which passed through the centre of a hollow windshaft. In front of the poll end, a cruciform iron 'spider' actuated bell cranks coupled to the shutter bars. The rear end of the striking rod terminated in a rack and this in turn was geared to a pulley or 'chain wheel' at the back of the mill. An endless chain, hanging down to the ground below, allowed the miller to move the striking rod backwards or forwards, and so operate the shutters. Automatic control was effected by hanging weights

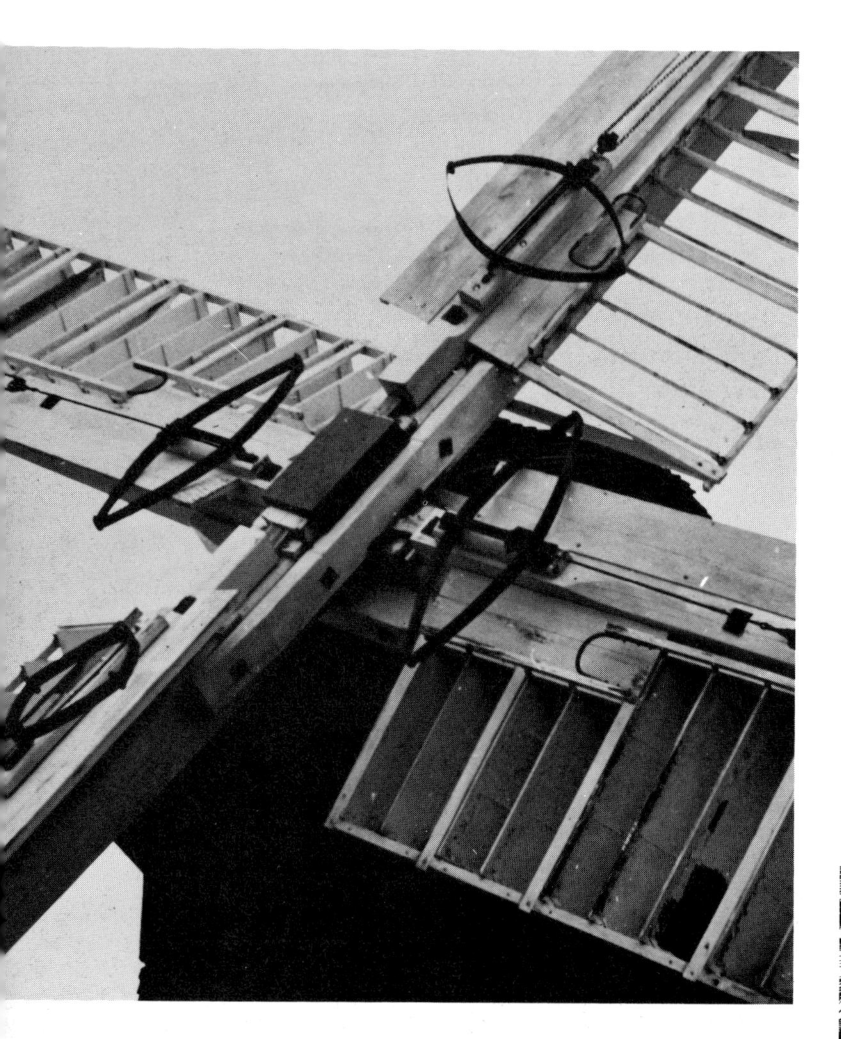

Shuttered sails controlled by springs at Outwood Mill, Surrey. The shutters here are in the 'open' position

A post mill, c. 1588, Ramelli: note the cloths laced through the sail bars. The quarterbars supporting the main post are doubled in accordance with common continental practice

ABOVE
An iron poll end at Horsev Mere, Norfolk

RIGHT
The miller takes in sail at The Swan, *a tower mill at Lienden, Holland*

LEFT
Patent sails at the Black Mill, Barham, Kent. The central 'spider' worked backwards and forwards, and actuated the shutter bars through bell cranks or 'triangles'

BELOW LEFT
A Lancashire tower mill equipped with roller reefing gear. Each sail is fitted with a series of canvas 'roller blinds' in place of shutters, actuated by the distinctive diagonal connecting rods or air poles

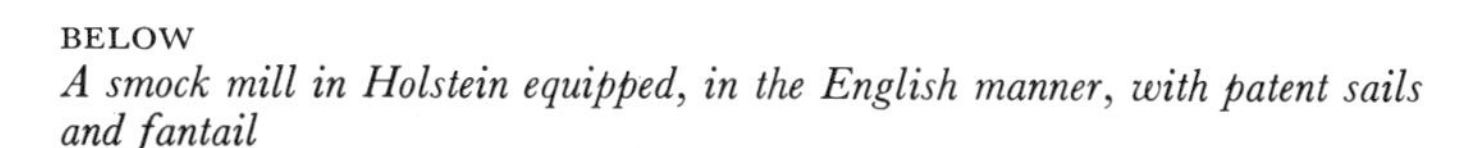

BELOW
A smock mill in Holstein equipped, in the English manner, with patent sails and fantail

on the chain, thus holding the shutters closed against the pressure of a working breeze. A somewhat similar linkage was used in the Berton sail, developed in France during the early nineteenth century. This was fitted with longitudinal shutters which could be controlled from within the mill, but the action was not automatic.

The nineteenth century also saw the introduction of the revolutionary annular sail, which consisted of a complete circle of vanes, arranged radially, in the form of a huge wheel. It was little used for mill work in Britain, but was widely adopted in America. Eventually, with the spread of the ubiquitous metal wind pump, it became a familar feature of the landscape throughout the world. The patent sail, however, enjoyed great popularity in Britain and its use spread to Denmark and Northern Germany, but the conservative Dutch millwright remained faithful to the common sail of earlier days. During the present century, much research has been carried out in Holland with the object of increasing the efficiency of the windmill. The sails of many working mills have been 'Dekkerised' by the addition of sheet metal fairings encasing the leading edge and stock, this streamlining materially increasing the power of the mill. Modern shuttered sails have also been developed, and it is interesting to see how closely the results of scientific research accord with the old designs reached empirically by British millwrights many years before.

All these developments took place in the North, but in Southern Europe the rate of change was very much slower, the primitive tower mills which survive in the central highlands of Spain having altered little since the first windmills were built in that region some 400 years ago. These mills are fitted with four sails of the symmetrical medieval form, but in Southern Spain and the Balearic Islands, a remarkable variation of the common sail assembly was developed. Here the typical tower mill was powered by six common sails which relied for their strength on three dimensional bracing. The windshaft terminated in a projecting spar, like a ship's bowsprit, from the end of which guy ropes radiated to the tip of each sail. Secondary guys, made fast to the hemlaths, maintained the 'weather' of the sail, while lines stretched around the perimeter, from whip to whip, braced each against its neighbours. Such a spider's web of rigging would have required constant attention

A windmill with an annular sail at Palatka, Florida, c. 1880

A tower mill at Zurrieq, Malta

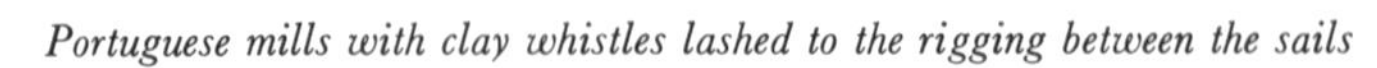

Portuguese mills with clay whistles lashed to the rigging between the sails

Multi-sailed post mills in Bessarabia, South-West Russia

The Maud Foster Mill, Boston, Lincolnshire. A number of multi-sailed mills were built in the Eastern counties of England, including several with eight sails. The sails were mounted on an iron cross in place of the conventional poll end

under the rain clouds of Northern Europe, but it served well enough in the Mediterranean sun. An advantage lay in the fact that it allowed the use of flimsy unwrought framing, an important economy in countries where good building timber was in short supply.

The most typical sail of the Mediterranean, however, was quite unlike the common sail of the North. The framing consisted of two sets of radiating spokes, mortised through the windshaft several feet apart. This double windwheel again derived its strength from bracing, the ends of the guy ropes being made fast to a boldly projecting bowsprit, but here the resemblance to the Majorcan mill ended. The sails themselves, up to twelve in number, were triangles of cloth, set, like the lateen sails of local fishing craft, on the forward spars, and sheeted home to the spars behind. Reefing was easily accomplished by wrapping the sail around its support until, in strong winds, only a minute area of white cloth would be visible at the yard arm. Sails of this type were used throughout the Aegean Islands, and in Greece, Crete, and Rhodes. Their use extended into the countries of Southern Europe bordering on the Black Sea, while far away to the West they became the characteristic sails of the squat tower mills of Portugal. Here wooden clappers or hollow clay whistles were lashed to the rigging, their mournful sound informing the farmer that the mill was at work, and warning him of changes in the wind.

In the working mill the cloths of the common sails were often tattered and patched. In Central Europe narrow longitudinal boards took their place, and were added or removed by the miller to suit the strength of the wind. Multi-sailed mills were not uncommon, although in the North, where the windmill achieved its highest peak of development, the cruciform sail assembly was always to remain the classic arrangement.

Whether the mill was in use or standing idle, the sails were kept facing into the wind. Not only was a true head wind necessary for efficient working, but the whole ponderous assembly of sails and windshaft was balanced to resist pressure from ahead, but not from astern. Like a sailing ship taken aback, a mill tail-winded in a storm was in grave danger, and the miller was always watching the weather, alert to any change in the direction or force of the wind. Turning a post mill into

The Swan—*the timber windlass in the cap*

BELOW RIGHT
The Swan *at Lienden, built in 1644, and fitted with inside winding gear. The sturdy cylindrical tower bespeaks an early date*

the wind was a comparatively simple operation, and the early tower mill was 'winded' in the same way by means of a tail pole, jutting sharply downward from the cap. But as towers grew taller, this ceased to provide a satisfactory lever arm and the miller found it increasingly difficult to manhandle the cap round from the ground below. Various types of 'winding' gear were developed over the years, and these were to have a profound effect on the appearance of the mill.

In the great Dutch smock mills, developed during the latter part of the sixteenth century, the tail pole was dispensed with altogether, and the cap was turned from within by means of a hand winch. A chain from the winch barrel was run out and made fast to a convenient point on the curb, and the cap was then wound round to face the wind. This process was slow and laborious, and few of these 'inside winders' now survive outside the provinces of Northern Holland. Millwrights in the South preferred to employ an improved and strengthened version of the simple tail pole, and this remains the normal practice in Holland. The pole was reinforced, and its leverage increased, by the addition of two pairs of raking braces, their ends bolted to transverse tie beams projecting on either side of the cap. A winch barrel was mounted on the lower end of the tail pole, from which a chain was hooked on to anchor posts set in the ground around the base of the tower. The winch was turned by a hand-wheel, often of considerable size to allow the miller to stand on the spokes, thus using his weight to turn the mill.

A notable advance was made in Britain, probably during the early eighteenth century, by the introduction of winding gear operated by an endless chain from the ground below. In its most usual form this gear consisted of a massive timber worm, cut from a solid baulk of hardwood and mounted horizontally below the rear gable of the cap. The worm meshed with wooden cogs projecting from the outer face of the curb, and was driven, through gearing, by a large wooden pulley wheel. By hauling down hand over hand on the chain, the miller was able to turn the cap in either direction with comparatively little effort.

Fully automatic luffing of the mill was to follow, made possible by the invention of the 'fantail' or 'fly', patented by Edmund Lee in 1745. This was a small wind wheel with wooden vanes, fitted to the back of

LEFT
Worm operated winding gear at Bembridge Mill, Isle of Wight: the endless chain has been removed

BELOW LEFT
The Rat *at Ylst—the braced tail pole and winch: note the brake lever projecting from the cap*

BELOW RIGHT
The winding gear of The Falcon, *Leiden: note the iron 'anchor' dropped between the timbers of the stage*

the cap, at right angles to the sails, and connected through gearing to the winding mechanism. As soon as the wind direction changed it would begin to revolve, slowly pulling the cap around to face the new quarter. The sensitive and restless little fantail, spinning silently to every shift and flaw in the wind, was quickly recognised as a valuable asset by the busy miller, and became a characteristic feature of the British mill. It was even adapted to the post mill, which it turned by means of a pair of iron truck wheels mounted in a carriage at the foot of the ladder. In Europe the fantail, like the patent sail, was adopted in those countries which followed the lead of the British millwright, but it never achieved popularity in Holland.

RIGHT
Winding gear at Cranbrook Mill: an inclined shaft links the fantail, through gearing, to the worm under the cap. Provision was also made for manual operation, an endless chain passing over the black 'Y' wheel and down to the stage below

BELOW RIGHT
A fantail geared to truck wheels at Holton post mill, Suffolk

BELOW
The fantail at Horsey Mere, Norfolk

The drive to truck wheels at Great Chishill Mill, Cambridgeshire

The brake wheel (left) meshing with the lantern pinion in a Dutch mill

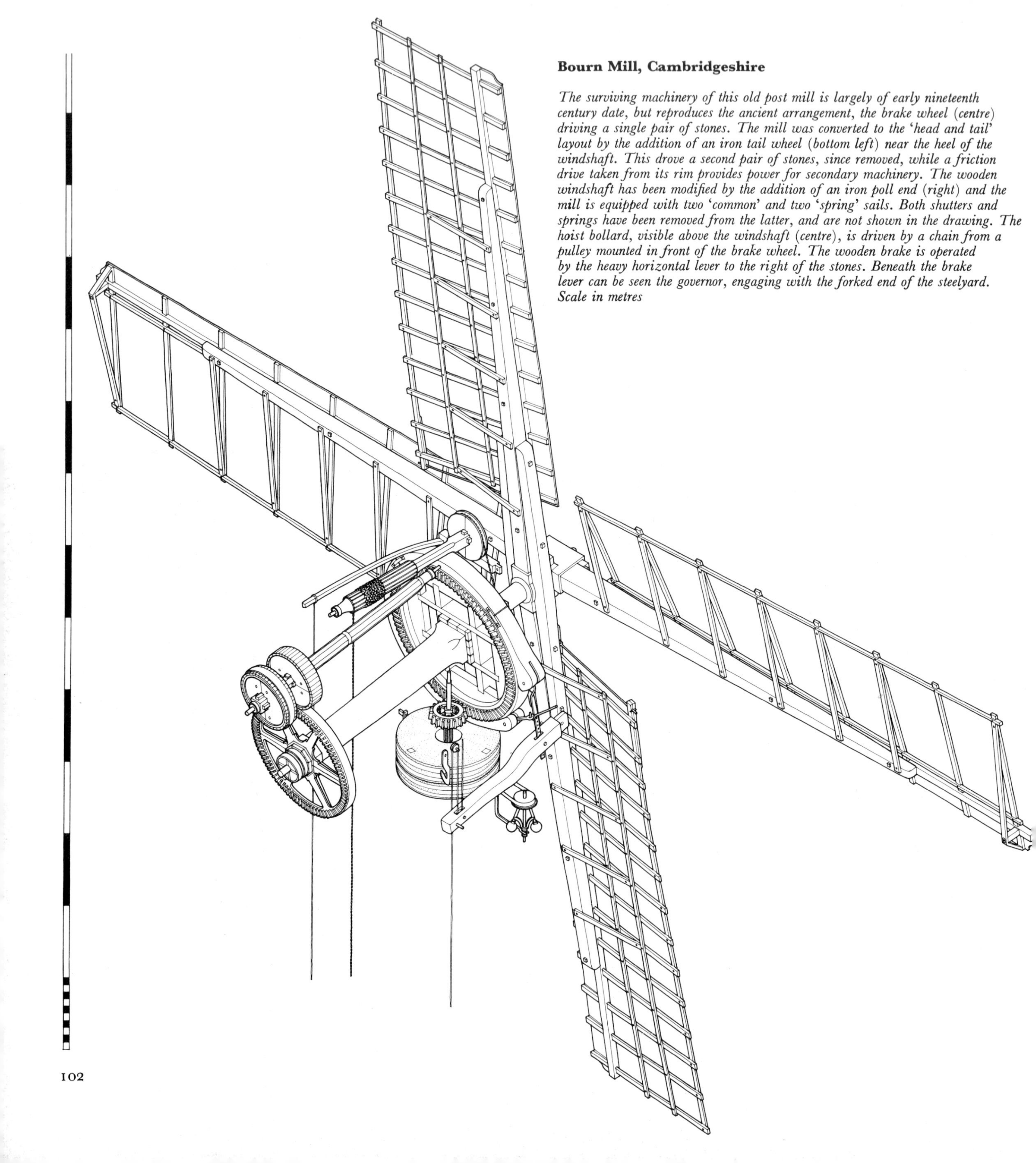

Bourn Mill, Cambridgeshire

The surviving machinery of this old post mill is largely of early nineteenth century date, but reproduces the ancient arrangement, the brake wheel (centre) driving a single pair of stones. The mill was converted to the 'head and tail' layout by the addition of an iron tail wheel (bottom left) near the heel of the windshaft. This drove a second pair of stones, since removed, while a friction drive taken from its rim provides power for secondary machinery. The wooden windshaft has been modified by the addition of an iron poll end (right) and the mill is equipped with two 'common' and two 'spring' sails. Both shutters and springs have been removed from the latter, and are not shown in the drawing. The hoist bollard, visible above the windshaft (centre), is driven by a chain from a pulley mounted in front of the brake wheel. The wooden brake is operated by the heavy horizontal lever to the right of the stones. Beneath the brake lever can be seen the governor, engaging with the forked end of the steelyard. Scale in metres

LEFT
The brake wheel at Stevington Mill, Bedfordshire

FAR LEFT
Bourn Mill—the clasp arm brake wheel mounted on the timber windshaft. The anchored end of the brake shoe can be seen bottom right, and the iron stone nut and quant bottom left. The top bearing or glut box has been removed

BELOW
The brake wheel and windshaft at Pitstone Mill

The machinery of the medieval windmill was modelled on that of the contemporary watermill, the basic differences stemming from the position of the horizontal drive shaft above the stones rather than below them. This arrangement was not universal, and in a curious type of Russian post mill, the stones were mounted above the windshaft, the necessary height being achieved by a tall underframe of unhewn logs. In general, however, power was transmitted downwards and the early mill was equipped with a single pair of stones, driven through a lantern pinion by a great timber face wheel mounted on the windshaft. In later days this became known as the 'brake wheel', from the massive band brake of elm or ash blocks fitted around its rim. There was a strong tradition among millers that early windmills were without brakes, and this no doubt consoled them for the shortcomings of their own, which were seldom really effective. Extreme caution was necessary when braking in a strong wind to avoid firing the mill. The brake was held against the rim of the wheel by a heavy timber lever or staff, and 'pulled off' by a rope passing over a sheave in the roof above and down through the floors of the mill. In tower and smock mills with oversailing caps, the brake line often hung down externally. In Holland a system of double levers was employed, the brightly painted operating pole jutting boldly out from the back of the cap and forming a most conspicuous feature of the mill.

In design and construction, the wooden gearing of the windmill passed through the same stages of development as the machinery of the watermill. Compass arm brake wheels were superseded by wheels of the clasp arm type, and the use of cast iron steadily increased during the latter half of the eighteenth century. This period saw the introduction of bevel gearing, to replace the traditional combination of face wheel and pinion. Many mills were fitted with composite brake wheels, their built-up rims of timber supported on cast iron 'spiders'. Others had brake wheels constructed entirely of iron, but traditional wooden gearing never passed wholly out of use even in Britain, while in Holland iron gearing remains to this day the exception rather than the rule.

Early windshafts were of timber, and surviving examples are often finely turned and chamfered. Friction at the neck journal presented a problem, and flush strips of iron were inset at this point to reduce wear. The bearing itself was formerly shaped from a block of stone or hardwood, but in later mills bearings of brass or gun metal were fitted. The cast iron windshaft, with its integral poll end, offered obvious advantages, and during the nineteenth century it was incorporated in new work wherever possible. A compromise was often effected by fitting a poll end of iron to an existing timber windshaft.

As post mills increased in size and power, a second pair of stones was installed in the after part of the buck, driven by a tail wheel fitted near the rear end of the windshaft. Old mills built to work a single pair of stones were often adapted to this 'head and tail' layout, and in such cases the facts are usually proclaimed by obvious differences in the age and workmanship of the machinery. On the continent, a different arrangement was customary, the brake wheel being cogged on both faces, to drive two pinions, one in front of the rim and one behind. It was not possible to place the two pairs of stones side by side on the centre line of the buck because of the limited distance between their driving spindles. They were therefore staggered, the pair nearer the tail of the mill being mounted at a higher level, an arrangement which had the advantage of keeping the stones well clear of the main post and crown tree.

Like the early post mill, the tower mill at first contained a single pair of stones, the runner being driven by a central vertical spindle. This

LEFT
A composite brake wheel and iron wallower at Chillenden Mill

BELOW LEFT
The brake staff at Chillenden Mill—to release the brake, the lever was raised until the iron pin (bottom centre) engaged with the swinging hook, above

TOP RIGHT
Neck bearings of stone and hardwood preserved in The Falcon, *Leiden*

CENTRE RIGHT
The tail wheel and wooden stone nut at Outwood Mill: note the miller's willow of steel (bottom right) tensioning the shoe against the quant

BOTTOM RIGHT
A wallower in the form of a lantern pinion at The Falcon, *Leiden*

BELOW
A tower mill equipped with a single pair of stones, c. 1588, Ramelli

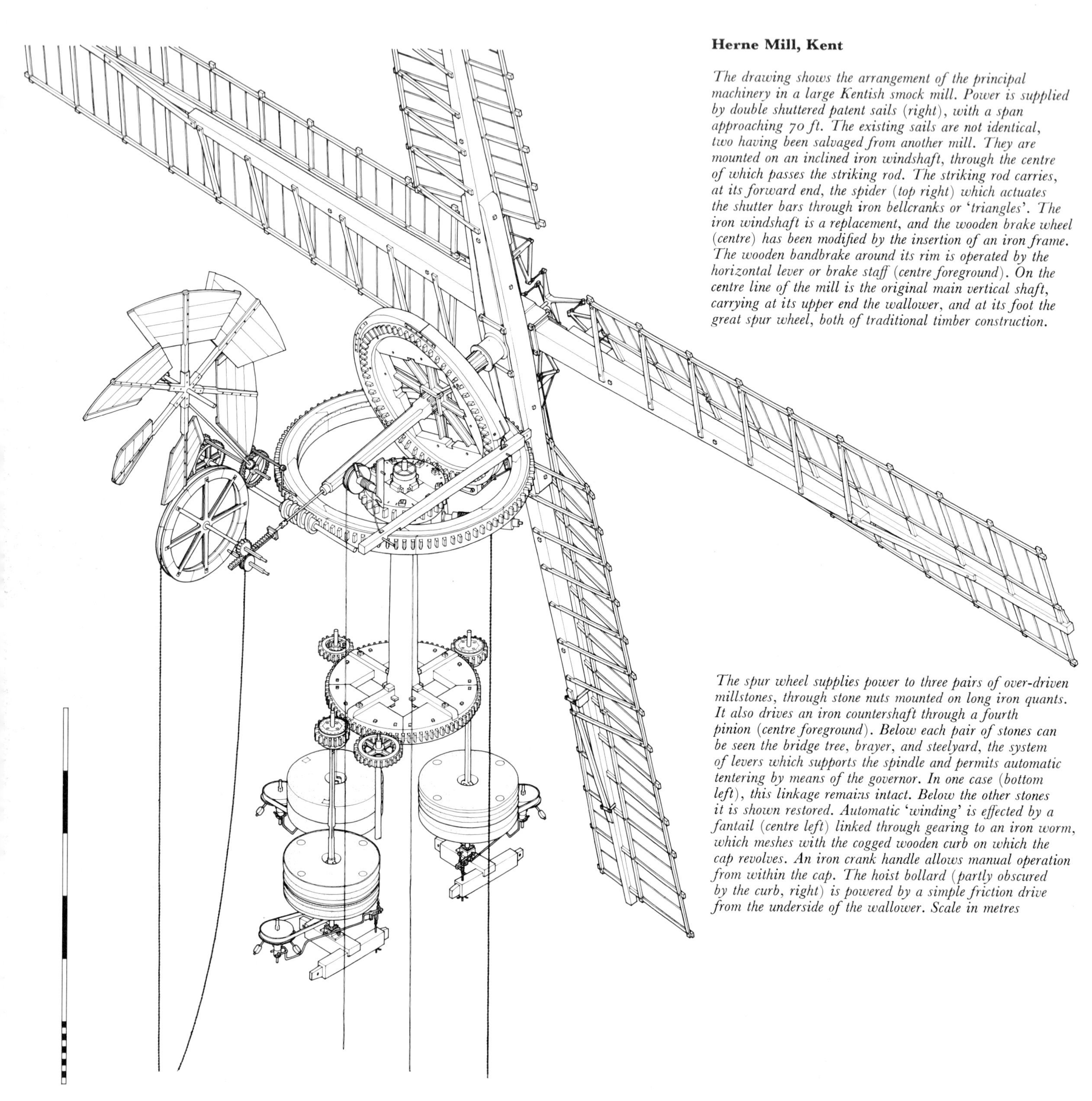

Herne Mill, Kent

The drawing shows the arrangement of the principal machinery in a large Kentish smock mill. Power is supplied by double shuttered patent sails (right), with a span approaching 70 ft. The existing sails are not identical, two having been salvaged from another mill. They are mounted on an inclined iron windshaft, through the centre of which passes the striking rod. The striking rod carries, at its forward end, the spider (top right) which actuates the shutter bars through iron bellcranks or 'triangles'. The iron windshaft is a replacement, and the wooden brake wheel (centre) has been modified by the insertion of an iron frame. The wooden bandbrake around its rim is operated by the horizontal lever or brake staff (centre foreground). On the centre line of the mill is the original main vertical shaft, carrying at its upper end the wallower, and at its foot the great spur wheel, both of traditional timber construction.

The spur wheel supplies power to three pairs of over-driven millstones, through stone nuts mounted on long iron quants. It also drives an iron countershaft through a fourth pinion (centre foreground). Below each pair of stones can be seen the bridge tree, brayer, and steelyard, the system of levers which supports the spindle and permits automatic tentering by means of the governor. In one case (bottom left), this linkage remains intact. Below the other stones it is shown restored. Automatic 'winding' is effected by a fantail (centre left) linked through gearing to an iron worm, which meshes with the cogged wooden curb on which the cap revolves. An iron crank handle allows manual operation from within the cap. The hoist bollard (partly obscured by the curb, right) is powered by a simple friction drive from the underside of the wallower. Scale in metres

The wallower at Lacey Green

Lacey Green smock mill, Buckinghamshire, built 1650—the primitive compass arm brake wheel (left) and wallower are rare survivals

The timber brake wheel and iron windshaft at Bembridge tower mill

Chillenden Mill, Kent, 1868

This is a late example of the open trestle post mill, powered by spring sails and equipped with two pairs of stones, mounted side by side in the breast and driven by spur gearing from a main vertical shaft. The brake wheel is of composite construction, but the windshaft and the bulk of the machinery is of iron. The stones are under-driven, mounted on iron bridge trees, and fitted with independent governors. Power for the auxiliary machinery is transmitted by an inclined shaft (left) from a face wheel mounted under the spur wheel. The mill, which was last at work in 1948, stands on an ancient site and incorporates timber from an earlier building. Scale in metres

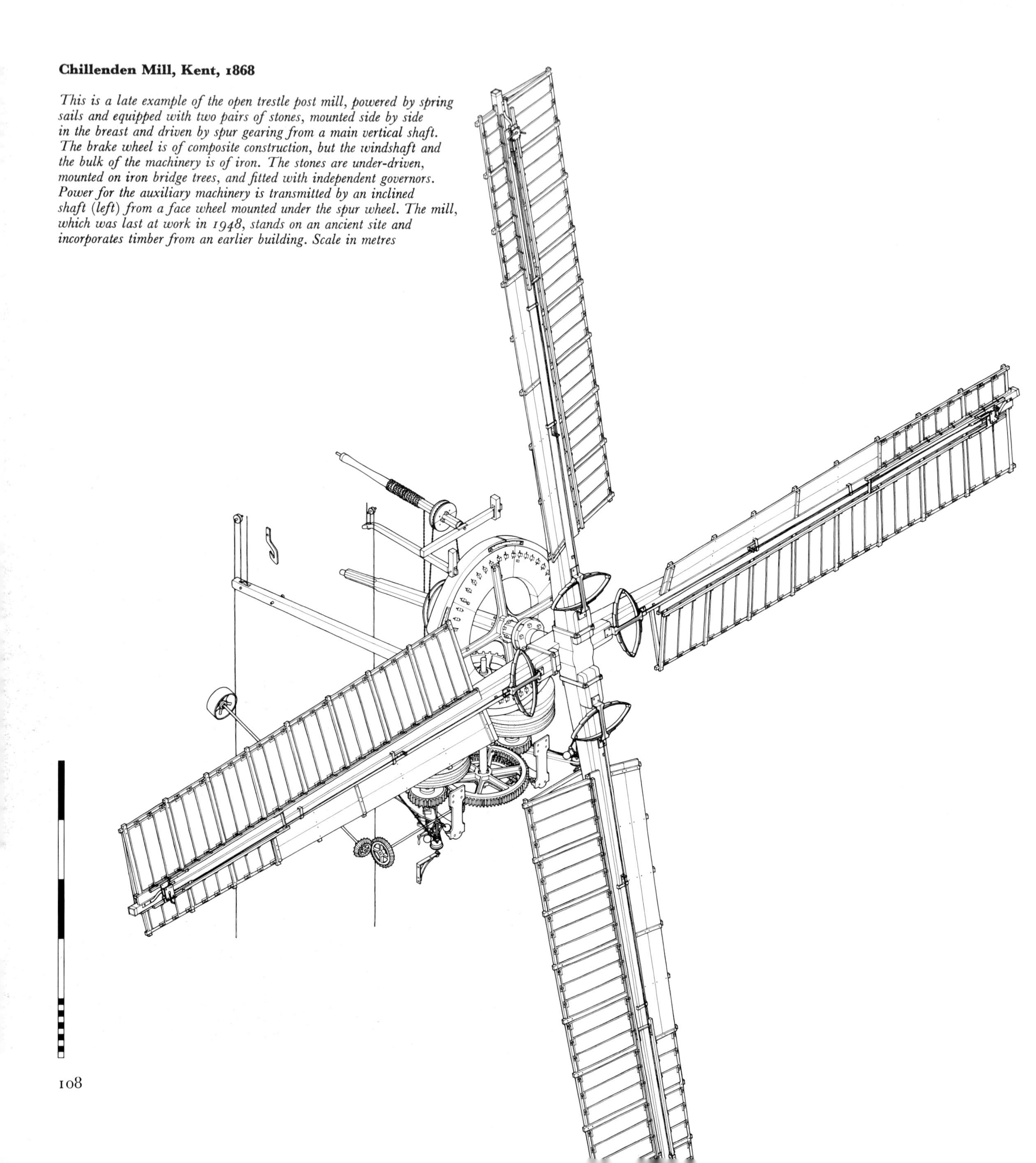

TOP
Herne Mill—the great spur wheel and foot of the main vertical shaft

ABOVE
The Falcon *at Leiden—a lantern pinion mounted on a timber quant*

RIGHT
Herne Mill—the spur wheel and wooden stone nut

Underdriven millstones side by side in the breast of Chillenden Mill—the runner stones have been removed

Bourn Mill—the governor. The underside of the bed stone is visible top left

Bembridge Mill—the crane was formerly in Wootton tide mill

arrangement worked well enough for many years, but difficulties arose when the millwright attempted to increase the capacity of the mill. The revolving cap made it impossible for him to adopt the methods used successfully in the post mill to drive a second pair of stones. The problem was solved by the introduction of spur gearing, which enabled two pairs of stones to be installed, one on each side of the central shaft. This innovation, which greatly increased the potential of the mill, was to have the most profound effect on the design of both wind- and watermills throughout the industrial countries of Northern Europe. Its use was to become so universal that in Britain no example of a tower mill fitted with a single pair of stones now remains. It was even adopted in the post mill, where it became customary to accommodate two pairs of stones side by side in the breast, in preference to the old head and tail layout.

The great spur wheel might be located either above or below the stones, which were said to be 'overdriven' or 'underdriven' accordingly. If placed below, the gearing resembled that of the watermill, and the stone nuts were disengaged by the removal of slip cogs. When iron pinions came into use, however, they were frequently keyed to the spindle so that they could be lifted bodily out of gear. Where the drive came from above, each stone nut was mounted on a 'quant', a long iron spindle terminating in a crutch which engaged with the rynd of the runner stone. To take a pair of stones out of gear, the miller would open the head bearing or 'glut box' and allow the stone nut to swing clear of the great spur wheel. Where this arrangement was adopted, the 'quant' also performed the duty of the damsel, flanges raised on its sides serving to shake grain down from the shoe into the eye of the runner stone below. The stones themselves, with their enclosing vats and wooden hoppers, were generally identical with those of the watermill.

An exception occured in the case of the hulling mill, once common in Holland, and equipped specifically for the preparation of barley or rice. Hulling was the process by which the hard outer husk was removed, and the stones were dressed with a few deep angular channels in place of the traditional layout of lands and furrows. These channels served to throw grain out to the circumference where it was grated between the edge of the stone and an abrasive outer casing of perforated metal.

In the water powered corn mill it was possible, by adjustment of the hatches, to achieve a more or less uniform speed, and thus ensure an even texture in the meal produced. The windmill, however, demanded a different solution. The force of the wind was constantly changing, and this had a curious effect on the action of the stones. As the speed of the mill increased, the runner stone would tend to rise on its spindle, opening the gap between the grinding surfaces, and making constant adjustment necessary in a choppy breeze. At first, this tentering was carried out by hand, the operation being eased either by the use of a pulley and counterweight, as in Holland, or by the introduction of a system of levers, to balance the weight of the runner stone. The end of the pivoting bridge tree was mounted on a secondary lever known as the 'bray', and this was in turn controlled by a long iron steelyard. In Britain the millwrights of the late eighteenth century developed a sophisticated method of automatic tentering by means of governors, which marked a great advance in technique, and was very generally adopted throughout the country. The governors employed were usually of the centrifugal type, and were driven by belts from the stone spindles or from the main shaft itself. As the machinery gathered speed the governor would raise the end of the steelyard, so lowering the bridge

LEFT
The Swan *at Lienden—the sack hoist driven by a friction wheel off the underside of the wallower (out of view to left)*

LEFT
The wooden governor at The Falcon, *Leiden*

RIGHT
Chain drive to the hoist at Chillenden Mill: note the roof of the buck curved to accommodate the brake wheel

A wooden countershaft driven from the great spur wheel at Bembridge Mill

tree and fractionally closing the gap between the stones. The required adjustment was very fine, and the tentering gear was most delicately balanced.

In the early Middle Ages the miller's customers carried their grain up the ladder and into the mill to be ground. The first sack hoists were probably no more than single whips, rove through a block or over a pulley projecting from the back of the mill, and such a device seems to be indicated in the fourteenth century brass to Adam de Walsoken at Kings Lynn. No doubt the hoist developed along the same lines as its counterpart in the watermill, and the location of the windshaft close up under the rafters must soon have suggested the use of mechanical power. In surviving British post mills the hoist usually takes the form of a horizontal shaft or bollard, carrying a wooden pulley at one end, and driven by a slipping belt, or chain, from a corresponding pulley on the windshaft below. The end of the bollard was raised by pulling on a rope hanging down through the floors of the mill, thus tensioning the belt and setting the hoist in motion. Methods of power transmission varied widely and the tail wheel was frequently employed, either to provide a direct friction drive or to operate an external hoist through spur gearing, as in many continental examples. In tower and smock mills with rotating caps, it was necessary to mount the hoist in the body of the mill, and in these cases the bollard was driven by means of a friction wheel brought into contact with the underside of the wallower.

The eighteenth and nineteenth centuries saw sweeping changes introduced into the traditional process of milling. This revolution, which has been briefly described in an earlier chapter, presented the wind miller with formidable problems, forcing him to install bolters and wire machines, separators and smutters. In the conventional post mill drives were commonly contrived from the brake wheel or tail wheel; in tower and smock mills equipped with spur gearing a countershaft provided the usual solution, transmitting power downwards through the lower floors. But in ancient mills the space available was strictly limited, and extraordinary ingenuity was displayed in accommodating and driving this array of new machines, unknown to the millwrights of earlier years.

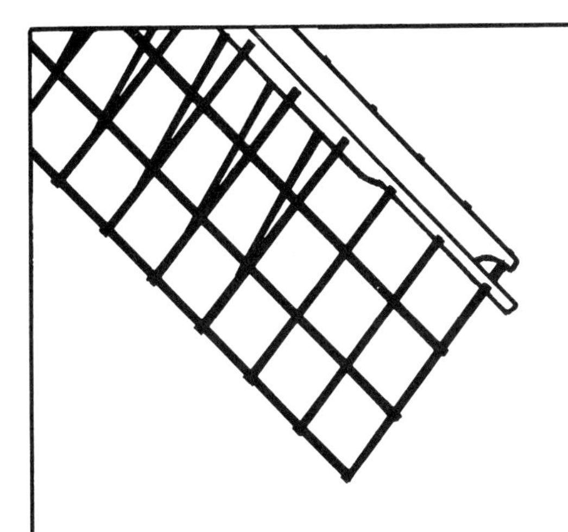

7 The Use of Water Power in the Textile Industry

Water power was first employed in the service of agriculture, either for irrigation, or for the processing of grain in the familiar corn mill. In Europe these were to remain the sole functions of the water wheel for nearly a thousand years, and it was only during the late tenth century that medieval man began to exploit the power of the running stream for other industrial purposes. There is a tendency to date the introduction of machinery from the period of the Industrial Revolution, and it is sometimes forgotten that this process had been set in motion long before. It is true that the eighteenth and nineteenth centuries saw a spectacular transformation of traditional crafts, with machines performing tasks which had been the jealously guarded mysteries of skilled tradesmen since time immemorial. But industries making limited use of machinery had been developing in Northern Europe throughout the Middle Ages, and the introduction of mechanical power for any aspect of a particular trade inevitably affected the time honoured pattern of production, paving the way for further development.

The production of woollen cloth involved at least six basic operations, some of which lent themselves more readily than others to the use of machinery. The first, and for centuries the only, process to employ water power was fulling. It is probable that the first fulling mills were set up in Italy during the closing years of the tenth century, and documents record the existence of others in Italy and elsewhere during the next hundred years.

During the early Middle Ages Britain was renowned for the export of raw wool rather than finished cloth. The fulling mill was evidently not known at the time of the Domesday survey, but it had become established by the end of the twelfth century. In succeeding years it was to play a vital part in the manufacture of woollen cloth, the country's greatest source of wealth.

'*Cloth that cometh fro the weaving is naught comeley to wear,*
Till it is fulled under foot, or in the fulling stocks . . .'

wrote William Langland in the closing years of the fourteenth century. The fuller scoured the freshly woven cloth, subjecting it to a pummelling action which permanently altered its texture by thickening and felting the fibres of the wool. Originally this was achieved by 'walking' the cloth, or by beating it manually, and these primitive methods survived in remote corners of the British Isles until the present century. A detergent was necessary to remove grease remaining in the wool and to wash out the size applied to strengthen the warp threads during weaving. Fullers' earth and later soft soap were the materials most usually employed, while urine was used as an alternative particularly in Wales and Yorkshire. The production of fullers' earth, a grey clay obtained principally in Surrey, formed a minor industry of such importance to the woollen trade that its export was banned in the seventeenth century. After thorough rinsing the cloth was dried in the open air, stretched on wooden racks known as tentering frames. Very considerable shrinkage took place during drying, producing a dimensionally stable material ready for the final stages of manufacture.

In the mill the beating process was performed mechanically in fulling stocks. The term 'stocks' originally referred to the container or fuller's pot in which the cloth was held but latterly it was used to describe the complete machine. Two basic types were employed, driving stocks and falling stocks. Driving stocks consisted of massive wooden mallets, their shafts supported at one end by trunnions and their heads raised and dropped by means of a tappet wheel. Falling stocks employed heavy wooden stamps working in a vertical plane, and operated by cams projecting from a rotating axle. In general, driving stocks were used where both scouring and felting were required, while falling stocks were more suited to the scouring process only.

Fulling stocks, c. 1617, Strada

In Britain, the first fulling mills were strongly opposed by craftsmen who saw the new machines as a threat to their livelihood. In London their use was forbidden in 1298, and sporadic opposition continued for centuries. In spite of this, the demand for fulling mills increased with the growing woollen industry, and this demand played no small part in deciding the geographical distribution of manufacturing centres. The amount of water power available in towns and cities was limited and often the number of mills was already excessive. The city of Winchester, for example, contained no less than twelve, and it is difficult to see how these can all have functioned efficiently. Such concentrations were possible with low-powered undershot wheels which simply turned in the current, but with the raising of weirs the problem became acute. In some cases existing grist mills were converted for fulling. Although such changes of use were fairly common, the demand for flour remained unaltered, and some mills served a dual purpose; thus Durngate at Winchester was used for both grinding and fulling during the sixteenth and seventeenth centuries.

The shortage of suitable sites led to the establishment of new fulling mills in rural districts where water power was readily available, and the town weavers were thus forced to carry their cloth out into the country to be fulled, possibly to the very district from which the yarn had originated. In the course of time, weavers settled around the new mill, which thus became the nucleus of a manufacturing community.

A small rural fulling mill still stands on the little River Alre in Hampshire. When the local woollen industry declined it was not converted to other uses, but remains substantially as it was built, the miller's house and mill sharing the same thatched roof. The present building appears to date from the seventeenth century but a fulling mill was recorded here in 1331. Power for the fulling stocks must have been obtained from a small undershot wheel, but the machinery no longer exists. An earlier mill, dating from the fifteenth century and known as 'the Abbot's fulling house', still stands at Glastonbury. Though much altered, it retains an original gable end with good stone detailing in the vernacular style. A few such mills, still concerned exclusively with the finishing processes for woven cloth, survived until recent years in Wales, where they were known as 'Pandys'. Some were little more than timber framed sheds enclosing the fulling stocks, indicating that, by contrast, the country mill may often have been a very humble affair. The word 'pandy' occurs frequently in the names of Welsh towns and villages, and in England too, place names recall this vanished industry. Many mills which ended their working lives grinding corn are nevertheless remembered locally as 'the fulling mill', or 'the tucking mill', while 'rack hill' and 'rack close' identify sites where tentering frames once stood.

In Wales a number of small woollen mills are still at work, and one has been happily preserved, in working order, in the grounds of the Welsh Folk Museum at St. Fagan's Castle. The Esgair Moel mill originally stood near Llanwrtyd in Brecknockshire, and worked commercially until 1947 to supply the needs of a rural community. The fulling stocks probably date from the first half of the nineteenth century, but their design is traditional. Here they perform only one of the many processes carried out under the same roof, but for hundreds of years they

The Abbot's Fulling House, Glastonbury, Somerset—the mill stream passes below the projecting wing (right)

The fulling mill at Alresford, Hampshire

The Esgair Moel Mill, St. Fagan's, Cardiff

The drawing shows the primary machinery of a small woollen mill. The fulling stocks (left) are driven by a tappet wheel mounted on the end of the main axle. In the drawing both hammer heads are secured in the 'raised' position. Their stepped profile is designed to turn the folded cloth during the fulling process. On the other side of the water wheel, an iron spur wheel drives the belt drum, from which power is transmitted to the machines on the floor above. Scale in metres

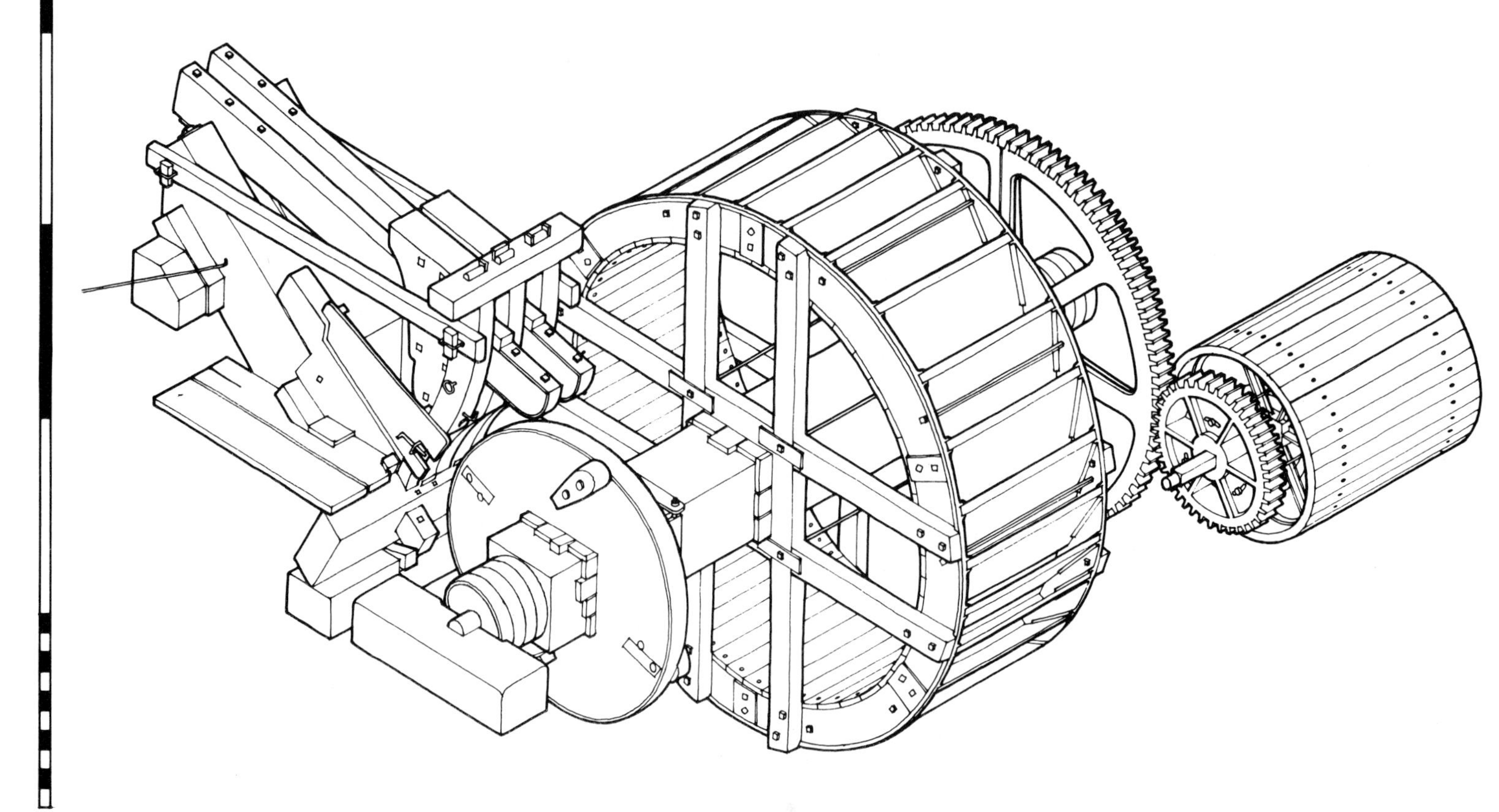

Esgair Moel Mill, St. Fagan's—a detail of the fulling stocks, showing the hammer heads (in shadow) raised clear of the tappet wheel (bottom right)

RIGHT
Model of a fulling mill above the gateway to the Lakenhal at Leiden, 1639. The mill is flanked by rolls of finished cloth

BELOW
Fulling stocks in the Moelwyn Mill, Blaenau Ffestiniog—the trip wheel is out of view to the left

were the only machines employed in the industry, and fulling was the principal function of the little mills in which they stood. In the Esgair Moel mill one is reminded of their former importance by the fact that the stocks alone are driven directly by the water wheel, and not by belting. The tappet wheel, a disc of solid timber 4 feet 6 inches in diameter and 3 inches thick, is secured by wedges to the ponderous square axle. As long as the water wheel turns the tappet wheel turns also, lifting and dropping the iron shod shafts of the hammers. To disengage them it is necessary to lift their heads clear of the tappets and retain them by an iron rod inserted through holes in the timber frame.

The overshot wheel is of composite construction, and the iron rims are cast in segments and bolted together. Where the fall was sufficient the overshot wheel was preferred by millwrights, being reckoned better able to deal with the intermittent heavy loading imposed by the hammers. In this case water is delivered to the wheel through a wooden launder, and the supply regulated by a cord hanging beside the stocks. A hinged section of the trough is lifted to direct water on to the wheel, or lowered to allow it to spill clear, while the removal of a bung diverts water to the stocks themselves.

The fuller required a plentiful supply of clean water, and wherever possible this was utilised as the source of power. Vertical water wheels were most commonly employed, and in districts where the primitive Norse mill held sway the ancient and laborious methods of fulling survived. In Holland, however, it was necessary to press the windmill into service. Enormous quantities of British wool were shipped across the Channel to feed the looms of Northern Europe, and many imposing cloth halls stand as memorials to this ancient industry. Such a building is the Lakenhal at Leiden, erected in 1639 by the cloth merchants of that city. The exterior is enriched with carved detail, recording the craft of the clothier, and in the place of honour above the entrance gate stands a fine bronze replica of a windmill. This faithful representation of a fulling mill of the period depicts a large smock mill, of the old type known in Holland as an 'inside winder', standing proudly between bales of cloth.

One of the last sets of stocks to work in the British Isles remains in the Moelwyn mill near Blaenau Ffestiniog, high in the mountains of

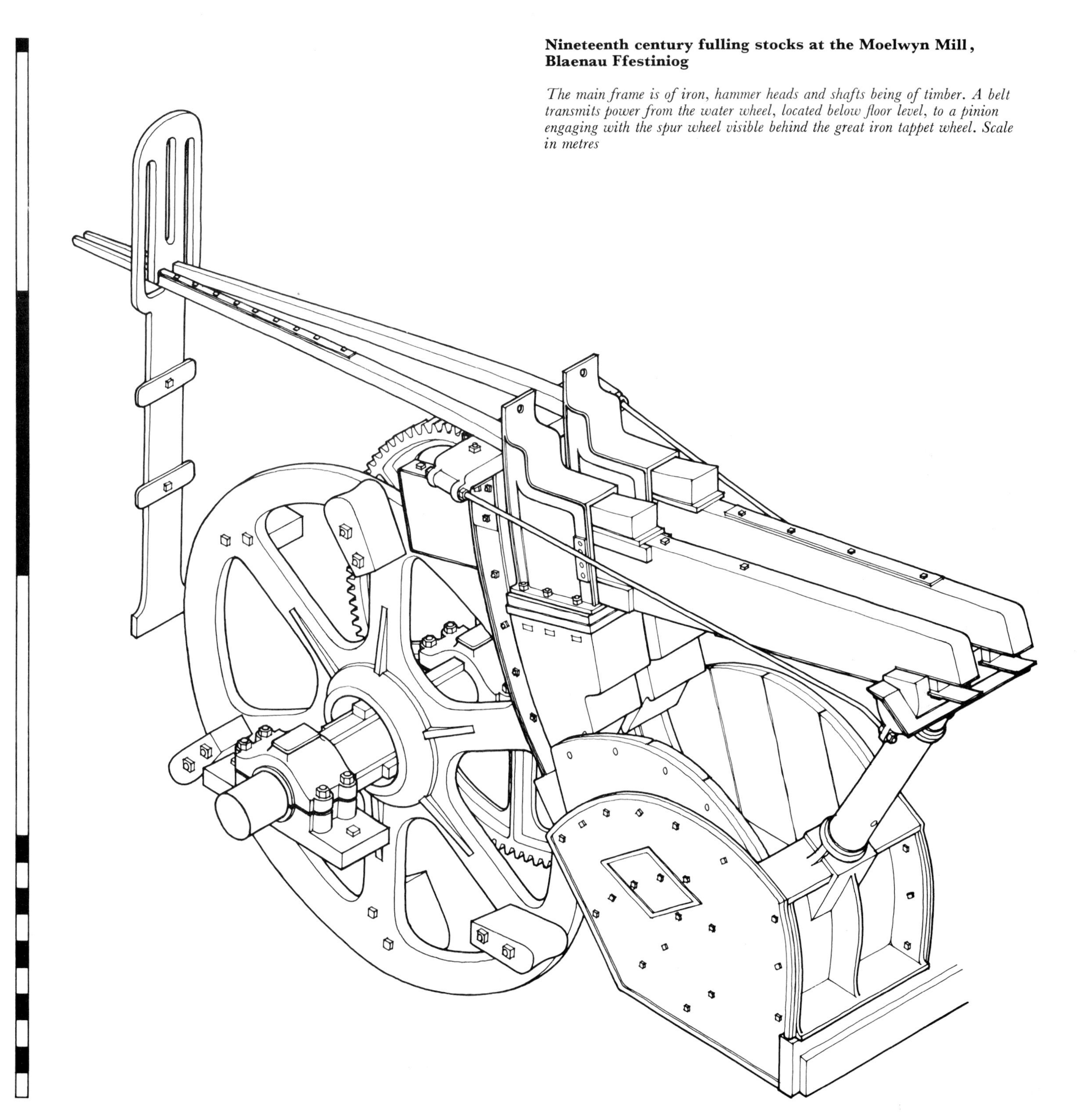

Nineteenth century fulling stocks at the Moelwyn Mill, Blaenau Ffestiniog

The main frame is of iron, hammer heads and shafts being of timber. A belt transmits power from the water wheel, located below floor level, to a pinion engaging with the spur wheel visible behind the great iron tappet wheel. Scale in metres

Merioneth. Again they performed only one of the many processes carried on in the building, and this is emphasised by the drive, no longer taken directly from the axle of the water wheel, but transmitted by belt and gearing. Cast iron here replaces timber in the construction of the frame, a stout inclined pillar supporting the hammer shafts. The body is built up from iron sections bolted together, and lined with hardwood bleached by years of use. The hammer heads and shafts are of oak, the heads being bolted to iron boxes secured to the shafts by wedges. The head of each hammer is carefully shaped, the steps and chamfers being designed to turn the cloth slightly with each stroke. The tappet wheel is a massive iron casting six feet in diameter, its lower half rotating in a pit formed in the floor of the mill. The hammers are too heavy to be lifted by hand, and they are disengaged and held out of gear by tapered wooden levers working in guides bolted to a partition behind the stocks. The miller recalls that the cloth to be fulled was carefully folded, while 'moist but not wet', and placed in the stocks with about a pound of 'clay'. A single loading might comprise a length equivalent to ten blankets.

The Moelwyn mill is entered directly from the street, and the thud of the fulling stocks must once have been a familiar sound in the village. They have now fallen silent, and a great baulk of oak lying in a corner of the mill will never be needed. This forlorn relic was cut to provide new hammer heads 'after fifteen years or so to season'.

A rotary milling machine, in which the cloth was passed through rollers, was invented by John Dyer of Trowbridge in 1833, and this method of fulling rapidly displaced the traditional process in the major manufacturing centres. The Welsh mills described remained in existence so long only because they worked in remote districts, supplying a local domestic market which had changed little during hundreds of years.

With the passage of time, machines similar to fulling stocks were introduced in other industries. Thus 'beatling' stocks, consisting of a battery of wooden stamps, performed the final process in the production of linen, imparting the characteristic gloss to the surface of the material. 'Stampers' were used in paper mills, and these are referred to in a later chapter, as are the related devices used in mining for the crushing of ore.

For many years, fulling was the only process in the production of woollen cloth to employ water power. Before the end of the fourteenth century, however, limited use was being made, in Italy, of a remarkable water driven machine for winding silk. This was the 'Fillatoio', an astonishingly ingenious contrivance consisting of a cylindrical frame carrying tiers of bobbins, operated by a revolving central shaft. But it was only in the latter part of the eighteenth century, and the early years of the nineteenth, with the inventions of Arkwright, Hargreaves, Crompton and their contemporaries, that any considerable progress was made towards the general mechanisation of the textile industry. Their machines led to the adoption of the factory system, and revolutionised the whole process of manufacture. But they were not designed specifically for the water wheel, and it was at this period that steam engines began increasingly to supplement the overburdened rivers as sources of power.

With few exceptions, the new machines were designed initially for the rapidly expanding cotton industry, and subsequently adapted for woollen cloth. One of the last processes to be satisfactorily mastered was that of weaving, and the power loom was slow in gaining universal acceptance. This was due in part to violent opposition from the weavers, and in part to doubts as to the efficiency of the new machines. Hand

RIGHT
Whitchurch silk mill, Hampshire

BELOW, FAR RIGHT AND BELOW RIGHT
Woollen mills near Stroud, Gloucestershire

loom weaving in fact never died out completely in the British Isles and is, of course, still widely practised in countries where the economy is based on agriculture rather than industry.

The fulling mill, therefore, did not give way immediately to the fully mechanised factory. During the intervening period, many new mills were erected, and existing ones modified, to combine fulling with 'scribbling', a preliminary carding process necessary to align the fibres of wool prior to spinning. A great many of the mills built during this period of transition were of considerable architectural merit. Early examples display some of the elegance found in domestic architecture of the period, while later mills often possess a quality of uncompromising strength, resulting from the functional use of load bearing brickwork and cast iron.

Whether designed for the production of woollen goods, canvas, or silk, these mills displayed basic similarities. In the little town of Whitchurch in Hampshire, beside the River Test, stands a handsome building of red brick, a working silk mill which retains many typical features. The front elevation is surmounted by a pediment, framing a clock installed to celebrate the victory at Waterloo, and the slate roof is crowned by an elegant wooden cupola. Similar mills were built in all the manufacturing areas, though the use of local building materials caused regional variations in character. Whitchurch mill is unusual in that it has survived unaltered and still produces fine quality silk from looms which worked by water power as late as the 1950s. From evidence remaining within the building it is possible to trace the way in which its use has changed since it was rebuilt, in its present form, during the late eighteenth century. Whitchurch stands near the fringe of the great wool producing area of Salisbury Plain, and the mill was built for the woollen trade. Of its original scribbling, carding and warping machinery little remains. Bulky equipment would have been installed on the ground floor, which is now occupied by power looms, the oldest dating from the late nineteenth century. Whether fulling was ever carried out in the building is not clear, and it is possible that this process was always performed in another mill which stands a short distance downstream. The upper floors are lit by windows at the front and rear, their iron frames still glazed with the original crown glass. Each floor is open from end to end, and unobstructed by partitions. Here the weaving was carried out on hand looms, and the sockets into which their uprights were set can still be seen in the floor boards. The frames of one or two looms happily survive, as do two spinning wheels. These relics illustrate the trend towards the establishment of factories where all the processes of manufacture could be carried on conveniently under one roof, even though some of them were still performed by hand, without benefit of water power.

Serge mill at Sticklepath, Devon

Whitchurch silk mill: the River Test divides to flow on either side of the mill

Whitchurch Silk Mill, Hampshire

Front elevation and section through the wheel house

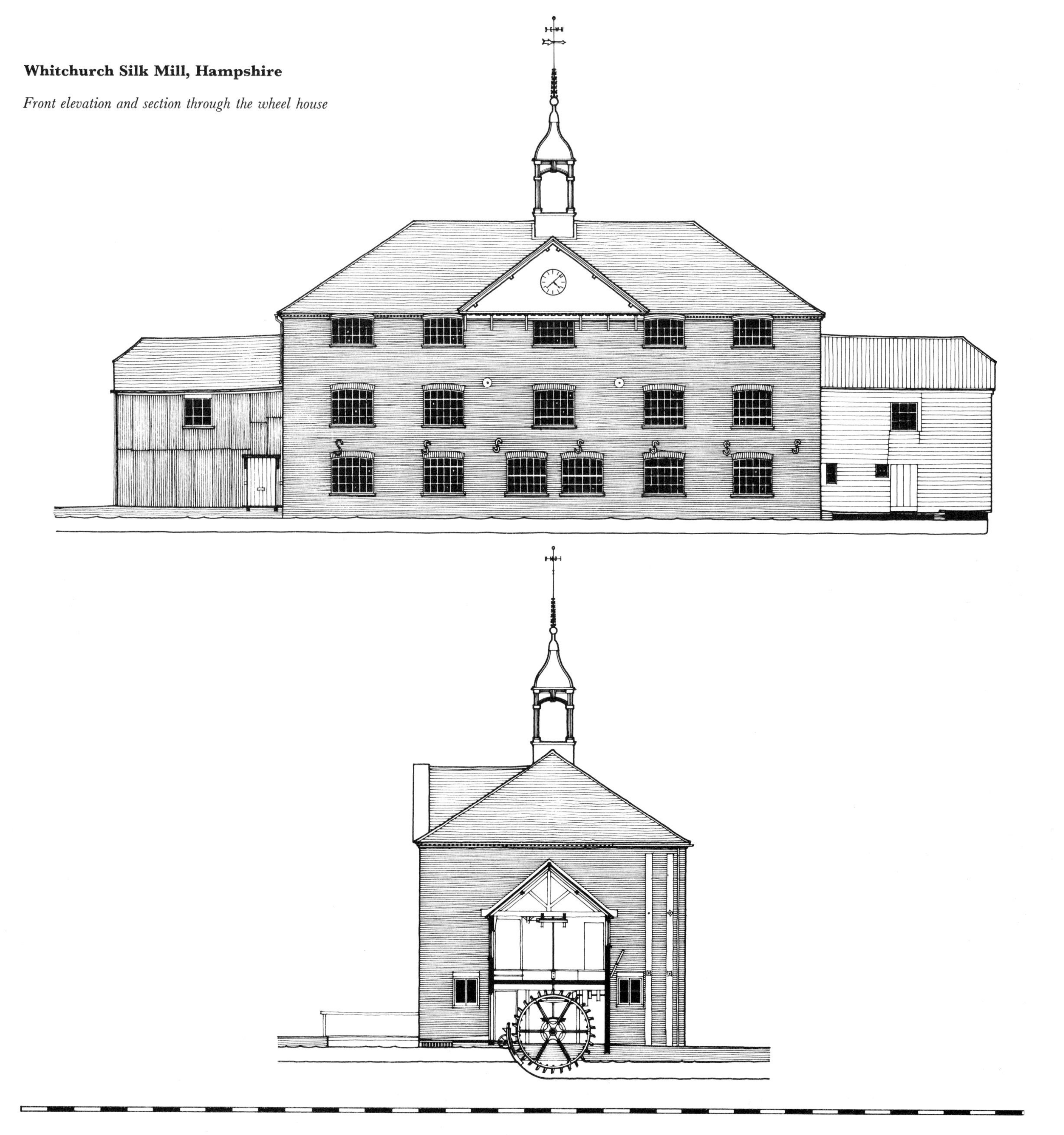

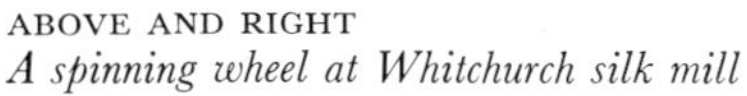

ABOVE AND RIGHT
A spinning wheel at Whitchurch silk mill

At Whitchurch, the water wheel and driving machinery remain intact. Each end wall of the mill is pierced by a wide opening in the brickwork, extending upwards through two storeys, and spanned by a massive timber lintel which supports the wall above. These openings give access to two timber framed wings, arranged symmetrically on either side of the central block. One of these extensions houses the water wheel, which is of low breast-shot type, with elm floats on a slender cast iron frame. Pit wheel, wallower and main shaft are all of iron, and the arrangement follows closely that of a corn mill of the period. In place of the great spur wheel, an iron face wheel five and a half feet in diameter once transmitted power to the machinery on the ground floor. The axle of the water wheel rotates in open bearings and the gear wheels are secured to the square main shaft by long iron keys. One is struck by the simplicity and elegance of the whole installation. The main vertical shaft is extended through an upper room to terminate in a bearing fixed beneath the tie beams. Just below the bearing is a second face wheel, this time constructed of timber with hardwood teeth, which supplied power, through a lay shaft, to the upper floors of the mill. A centrifugal governor, connected by a sophisticated train of rods and gears to a clutch mechanism, controlled the speed of the water wheel by raising and lowering the hatch. Machinery throughout the building was formerly driven by belts and pulleys from lay shafts bolted to the underside of the beams. When the water wheel was finally abandoned these shafts were dismantled, and the looms are now operated by individual electric motors.

The evils of the factory system during the nineteenth century are well known and indisputable. But the study of a building such as Whitchurch Mill gives a glimpse of the other side of the coin. Silk mills were widely considered to bring great benefit to towns by their demand for unskilled labour, and advertisements in contemporary papers extol their virtues to parents and guardians of the young. Whitchurch Mill employed 108 persons in 1838, including 39 children under 13 years of age. They worked hard and for long hours, but the view from the windows, then as now, was over green water meadows. The sound of the race could be heard in every corner of the mill, and trout must have hung, as they do today, in the clear green water under the floor of the wheel house. One suspects that the manager at least would fish when he had the time; a fly rod hangs now between two of the windows on the first floor. It is easy, when the mill has stopped work for the day, to imagine the presence of the children who worked there. The random tapping of driftwood against the hatches gives an illusion of life to the mill, and one looks round involuntarily for the child who scratched a picture of a house, with sash windows and smoking chimney, on the frame of a loom a hundred years ago.

Whitchurch Silk Mill, the principal machinery

The basic layout closely resembles that of the traditional corn mill. Power was transmitted through the building by lay shafts mounted beneath the ceilings. The governor (bottom centre) is linked to an intricate clutch mechanism designed to raise or lower the hatch, and thus control the speed of the machinery. Scale in metres

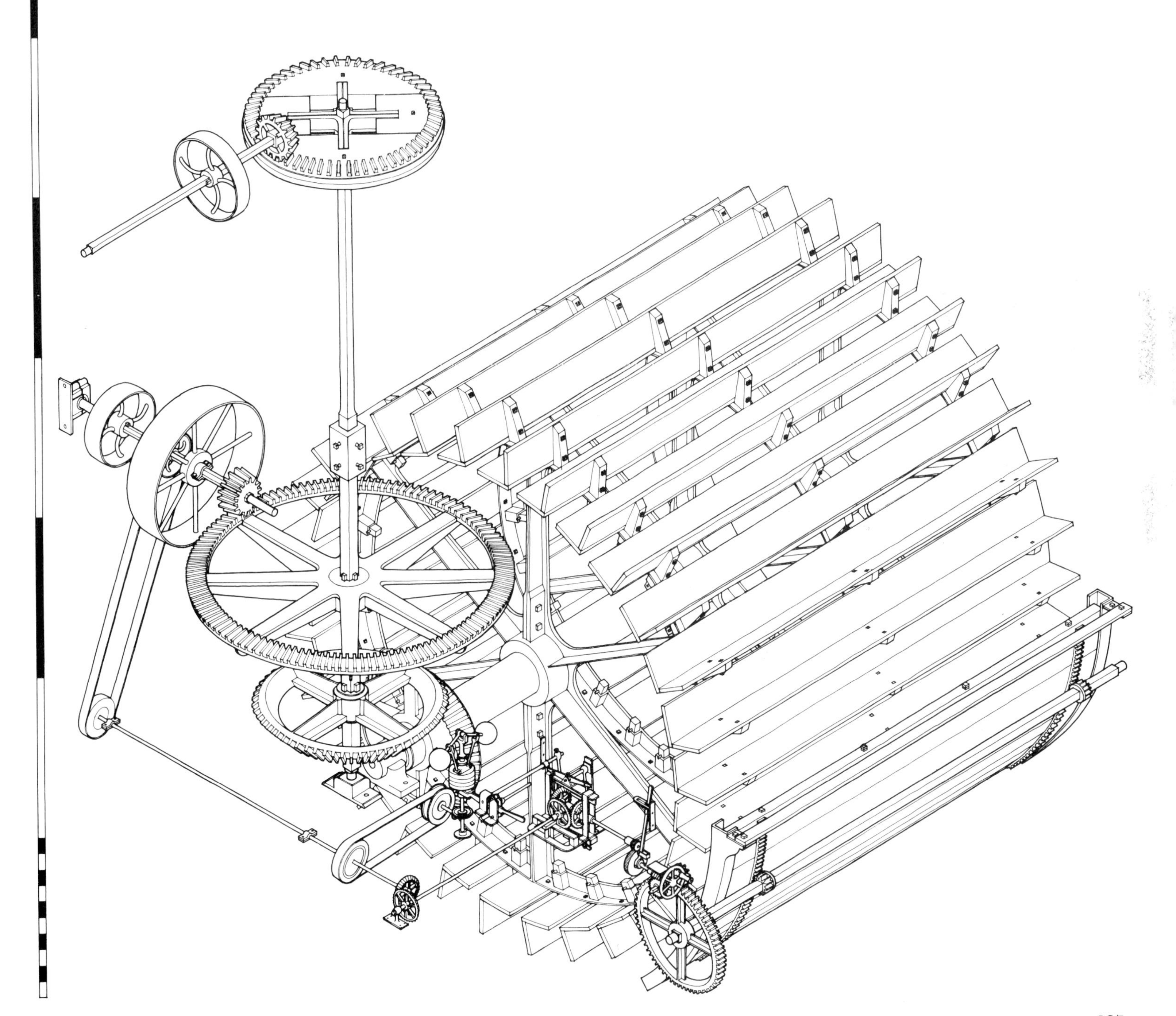

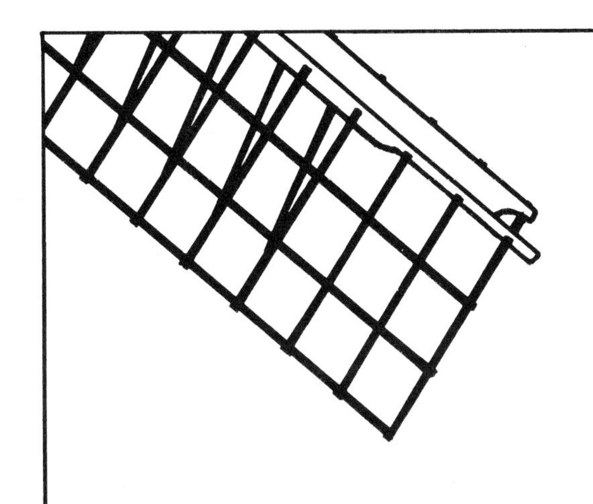

8 Water Powered Drainage Mills

Wherever climatic conditions made irrigation necessary, and streams were available to provide motive power, the primitive wheel of pots might once be found, turning slowly in the current as it lifted precious water to the parched fields along the river bank. It is now being supplanted by more modern machinery, but may still be seen at work in countries as far apart as Portugal and China. This ancient device, from which the vertical wheel of the Roman mill was derived, consisted of a tall, lightly built undershot wheel of timber with jars of earthenware lashed around its rim. Its function was to scoop up water from the river, and discharge it into a wooden trough projecting from the bank above.

The old city of Hama in Syria (the biblical *Hamath*) was long supplied with water from the River Orontes by enormous wheels. These, together with the towering stone aqueducts which they served, were visible from a great distance, and were for centuries a source of wonder to travellers. In Europe, the irrigation wheel, or noria, remains in use in Portugal, particularly along the River Mondego. Crude groins of brushwood are erected across the shallow bed of the stream to channel water to the wheels, which may measure as much as 40 feet in diameter. The numbers of jars are adjusted to suit the strength of the current, and the wheels are then left to work on untended throughout the day.

In China, hollow bamboo tubes were used as water containers in place of clay pots, and several variants of the simple noria were developed for use on sloping river banks. Some of these employed endless chains of bailers, while others were in fact primitive chain pumps with rectangular wooden pallets, drawing water up inclined tubes of timber.

Northern Europe had little need of the irrigation wheel, and the efforts of engineers were directed towards ridding the land of surplus water rather than conserving it. During the late medieval period the water wheel was brought into service for mine drainage on the continent, and in England it was in use for this purpose by the close of the sixteenth century. The machinery employed in Saxony at this early period was faithfully recorded in a series of detailed drawings by the German physician, Georgius Agricola, whose *De Re Metallica* provides an unparalleled source of information on early metallurgy and mining techniques. His illustrations show large overshot water wheels of both compass and clasp arm type, their massive timber axles revolving on

A wheel of pots at Hama, Syria

Mine drainage by bucket and windlass in Saxony—the overshot wheel illustrated is reversible; from De Re Metallica *by Georgius Agricola (Georg Baur), 1556*

A—Reservoir. B—Race. C, D—Levers. E, F—Troughs under the water gates. G, H—Double rows of buckets. I—Axle. K—Larger drum. L—Drawing-chain. M—Bag. N—Hanging cage. O—Man who directs the machine. P, Q—Men emptying bags.

iron gudgeons. Water was raised by various methods, the simplest of which employed the axle of the wheel as a winding drum, leather buckets being drawn up on a chain and emptied by hand at the top of the shaft. This installation was powered by a reversible wheel, fitted with two sets of opposed buckets, and controlled by a winch man perched in a wooden gallery above the mine shaft.

Other machines illustrated were of far more advanced design. For relatively shallow workings, reciprocating pumps were used to lift the water in stages from the bottom of the shaft. The pump barrels, consisting of straight tree trunks bored out with augers, were in some cases arranged in pairs so that a reasonably constant load was imposed on the driving mechanism. At the top of its stroke each pump discharged into a horizontal trough or reservoir from which the lift was continued towards the surface. The pump rods were of iron and carried perforated plungers of wood or metal fitted with leather valves. Power was transmitted from the axle of the water wheel by a stout iron crank, an early instance of the use of this device for the conversion of rotative to reciprocating motion.

Suction pumps powered by an undershot wheel, Agricola

The most typical and widely used machine, however, was the rag and chain pump. This consisted of an endless chain, threaded through a continuous pump barrel of hollowed logs, and provided with pallets of leather stuffed with rags. The chain passed over a horizontal axle on which was mounted a driving wheel, part pulley and part sprocket. In practice these chain pumps achieved remarkable results with lifts of over 200 feet being recorded. Their great advantage lay in the use of continuous rotary motion and they were also relatively simple to construct and maintain. A drawing dating from the late seventeenth century shows a pump of this type, driven by two overshot wheels, in use at Mostyn in Flintshire, and although more sophisticated machines took over many of the duties of the chain pump in later years, it was to continue in limited use into the twentieth century.

The rapid improvement in metal working techniques during the early years of the eighteenth century paved the way for the construction of larger and more powerful lift and force pumps for mine drainage. Flooding presented a particular problem in the mines of Cornwall, and over the years Cornish engineers were to earn an international reputation for the design of pumping machinery. Power was transmitted from the water wheel to the pump rod by a massive rocking beam of timber. This characteristic linkage was later adapted to the service of steam, and the general form of the beam engine, most monumental of all steam driven machines, was thus derived directly from the water powered pump. One may, perhaps, trace the influence of these early engines still further, and ascribe to them some responsibility for the towering engine houses which must rank amongst the most dramatic architectural features of early industrial sites.

The first experiments with an alternative source of power had been made in England during the closing years of the seventeenth century by Thomas Savery. His 'fire engine', an atmospheric pumping device of curiously organic appearance, was used with some success in the drainage of mines in Devon and Cornwall. It was superseded by the atmospheric engine of Thomas Newcomen, in which the partial vacuum produced by the condensation of steam was utilised to produce recipro-

A rag and chain pump powered by an overshot wheel, Agricola. The pump barrel, with a sheave at the foot, is shown in detail (right)

A piston forcing pump serving a water tower, c. 1588, Ramelli. The basic drive of the watermill is here adapted to the operation of pumps

cating motion. The erratic stroke of the Newcomen engine, which at first made it unsuitable for driving rotative machinery, had little adverse effect on its use for pumping, and it was widely employed for this work. It is a curious fact that the first steam engines installed in mills and factories were pumps. They did not drive machinery directly, their function being simply to maintain an adequate head of water in the mill race above the wheel.

In the early years of its development, the steam engine was regarded in the industrial field as a rather expensive and unreliable substitute for water power. As late as the middle of the nineteenth century mines were provided with water driven pumps whenever a suitable supply could be exploited, and many of these installations were designed on a grand scale. Huge overshot wheels were set up where falls allowed, and considerable engineering works were undertaken in the construction of leats and aqueducts. In some cases the wheel could not be located close to the mine shaft and it was then necessary to transmit the drive to the pumps by means of flat rods working over pulleys. At the Devon Great Consols Mine these rods extended uphill to the shaft for a distance of no less than half a mile. Here the wheel measured 40 feet in diameter, and turned at four revolutions per minute on a gigantic axle of timber. This spectacular engine was famous in its day and attracted numbers of visitors.

The largest pumping wheel ever built survives at Laxey in the Isle of Man. Measuring more than 70 feet in diameter, and weighing 100 tons, the mighty *Isabella* wheel must surely rank as one of the most impressive engineering works of the Victorian era. Built in 1854, it forms the centre-piece of an extraordinary architectural composition, almost Roman in scale and feeling. The bearings are mounted on a massive substructure of stone, linked by a causeway to the shaft of the Great Laxey lead mine further up the valley. Power was transmitted through a crank on the main axle to a heavy horizontal connecting rod, mounted on a rocker arm. The pump rod itself, which extended to the bottom of the shaft, was 1500 feet in length, and its enormous weight was counterbalanced by two bobs. Water was delivered under pressure, through a tall tapering column of masonry and along a wooden flume. The wheel turned at two and a half revolutions per minute, and remained in service until the workings were abandoned in 1929.

As the great cities of Europe grew in size, the provision of adequate supplies of unpolluted water presented an increasing problem. John Bate, whose book *Mysteries of Nature and Art* was published 1634, described how the needs of London were met, in part at least, by 'a water-mill or engine near the North end of London bridge'. Built by a Dutchman, Peter Morris, and completed in 1582, this pump was driven by an undershot wheel. The water was raised into a 'turret' or water tower and was distributed by gravity to an area of the city 'above two miles in compass'. During the period of the Commonwealth, means were again sought to improve London's water supply and the task was undertaken by Edward Ford, a country gentleman from Sussex. According to a contemporary account, Ford installed a new wheel coupled to 'a rare engine of his own design', which in 1656, 'raised Thames water into all the streets of the city'. No traces of his work remain in London, but at South Harting in Sussex may be seen the mill pond and pump house which he built to serve his own estate.

By the end of the seventeenth century some very large pumping installations had been constructed. The water works at Marley on the Seine, built in 1682, employed no less than 13 water wheels arranged in a barrage across the stream and driving a complicated system of pumps. The machinery at London Bridge was renewed on an impressive scale

BELOW
The Great Wheel at Laxey, Isle of Man, 1854

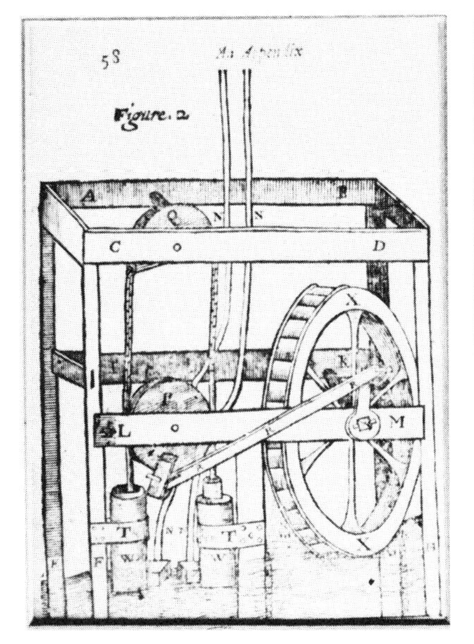

FAR LEFT

The pumping engine installed at London Bridge in 1582 by Peter Morris, from Mysteries of Nature and Art *by John Bate, 1654*

LEFT

The pump house at South Harting, Sussex, seventeenth century: built to serve the mansion of Uppark, perched on the hillside 350 feet above the village

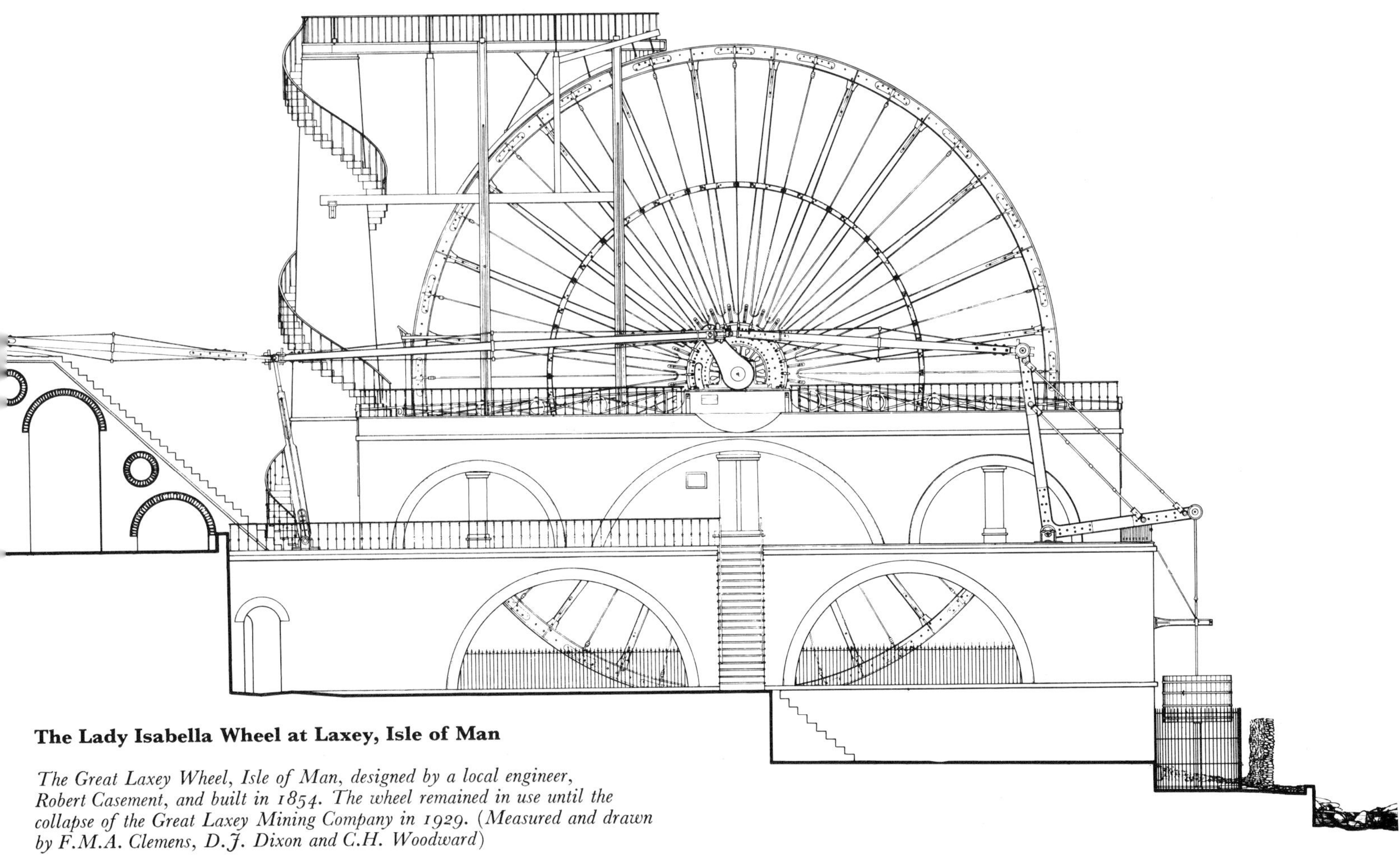

The Lady Isabella Wheel at Laxey, Isle of Man

The Great Laxey Wheel, Isle of Man, designed by a local engineer, Robert Casement, and built in 1854. The wheel remained in use until the collapse of the Great Laxey Mining Company in 1929. (Measured and drawn by F.M.A. Clemens, D.J. Dixon and C.H. Woodward)

LEFT
Water works at Marley on the Seine, 1682—built by Louis XIV to supply the water gardens at Versailles, from Theatrum Machinarum Molarium, *by Jacob Leupold, 1735*

BELOW
George Sorocold's pumping engine at London Bridge, early eighteenth century: Universal Magazine, 1749

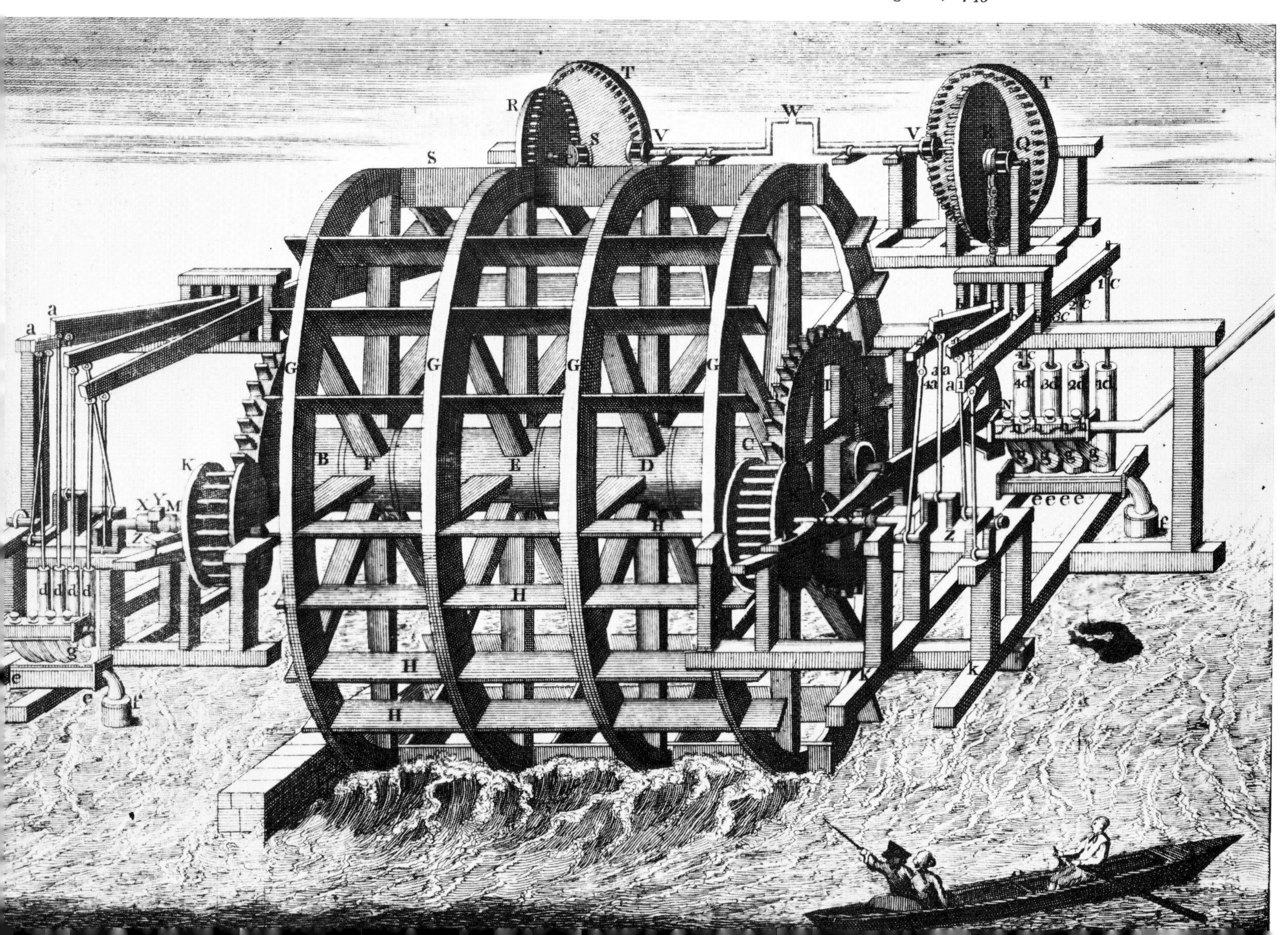

A breast-shot wheel from a domestic pumping engine, late nineteenth century

in the early eighteenth century by George Sorocold of Derby. His great undershot wheel measured 20 feet in diameter by 14 feet in the breast, and it drove through cranks and connecting rods a total of 16 pumps. The main bearings were mounted on hinged beams, which allowed limited vertical adjustment to suit the state of the tide, thus overcoming one of the principal disadvantages of the undershot wheel.

During the eighteenth century, improvements in the design and construction of pumping machinery were of the greatest importance to the general progress of mechanical engineering. The speed of advance may be gauged by comparing Sorocold's machine with a new pump, set up at London Bridge some 50 years later, and shown in an engraving of 1768. Many country houses were equipped with pumping engines, and by the end of the nineteenth century various compact and self-contained machines were available for domestic use. Diminutive breast-shot wheels, often measuring no more than three feet in diameter, were mounted in iron troughs and coupled to small horizontal pumps. Some few of these attractive little machines, brightly painted in the colours normally associated with agricultural implements, may still be found in isolated farm houses.

A notable example of an early pumping engine survives at Melin Griffith on the northern outskirts of Cardiff. It has outlived the canal that it was built to serve, and now stands half hidden among trees and nettles, a mute reminder of past industrial activity. An even more remarkable installation is to be seen at Claverton in Somerset, near the western end of the Kennett and Avon canal. Here a pumping station was set up in 1813 to supplement the natural supply of water. The site selected was at a point where river and canal ran close together, the Avon in its steep wooded valley, the canal hugging the contours of the hillside above. The river already supported a full complement of mills, and one of these was bought by the Canal Company to secure a power supply. The weir was raised, in spite of opposition from neighbouring millers, and the old building demolished to make way for a new pump house. It is a sober, dignified building of grey stone, with the round headed windows and classical detail characteristic of the period. The timber wheel house encloses a great breast-shot wheel of composite construction, with wooden floats bolted to a cast iron frame. The wheel

A pumping engine installed at London Bridge in 1768

RIGHT
The pump house at Claverton, Somerset, serving the Kennett and Avon canal, 1813. The canal passes along the side of the valley to the left

BELOW
Water powered pumps at Melin Griffith, Glamorgan: the water wheel is visible in the background

measures 19 feet 4 inches in diameter by 25 feet in the breast. In practice this width was found to be excessive, and after various repairs, the wheel was rebuilt in its present form incorporating a central bearing to reduce the strain on the axle. The speed of rotation was normally controlled by adjustment of the sluices, but provision was made for throwing the pit wheel out of gear in any emergency which might arise through failure of the hatches, or a sudden spate of flood water.

Within the engine house a pit wheel transmits the drive, through gearing, to a pair of cranks and connecting rods, their motion regulated by a five-ton fly wheel, an unusual feature in a watermill. The mechanism follows the familiar pattern of the beam engine, but with the direction of the drive reversed. An upper floor allows access to the beams themselves, which are of cast iron, and to the parallel motion linking them to the pump rods. The two 19 inch bucket pumps drew water from the mill race and discharged it through a cast iron delivery pipe into the canal above. The vertical lift amounted to 53 feet, and at 16 strokes per minute the pumps were able to raise 77,000 gallons per hour.

In 1852, the Canal Company was bought out by its rival, the Great Western Railway, and a line was driven along the side of the valley, within feet of the pump house wall. Traffic on the canal declined, but through all these changes the pumps laboured on, until finally halted by a comparatively minor breakdown in 1953. They had worked continuously for 140 years, with only brief stops for servicing and repair, and their extraordinary record of service is a testimony to the engineers who built them.

Electric locomotives now thunder past the windows which once reflected only the barge horses plodding by along the towpath above. But inside the silent building little has changed. Sunlight streaming through the dusty windows picks out the wooden patterns from which the gear wheels were cast, still hanging in place on the walls. It strikes through into the dark recesses of the wheel house to show the great bulk of the wheel, standing idle at last behind its closed hatches. There is at Claverton that strange air of suspended animation, so often to be sensed in old mills, which results from years of patient maintenance rather than conscious preservation.

LEFT
The Claverton pumps: note the timber roof trusses arched over the huge breast-shot wheel

BELOW LEFT
The Claverton pumps—the water wheel viewed through the spokes of the flywheel

BELOW
The Claverton pumps—the pit wheel (left) and the flywheel with connecting rod in the centre background

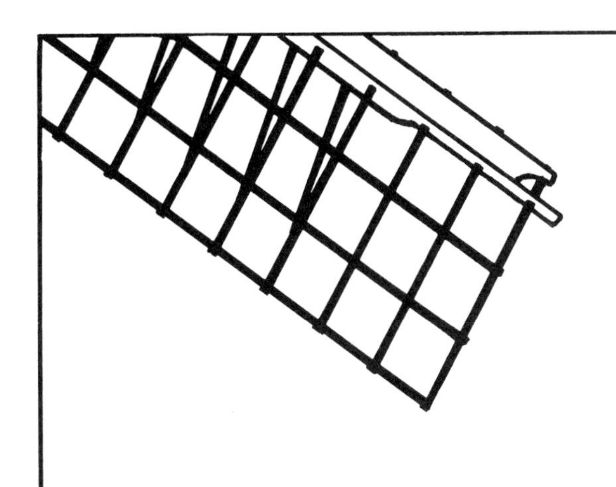

9 Wind Powered Drainage Mills

A hollow post mill, or wip mill, at Groot Ammers near Schoonhoven, Holland

'God made other countries—the Dutch made Holland for themselves'—with this proverb the Dutchman was wont to justify his national pride, and it is true that his ceaseless efforts over the years have resulted in the reclamation of huge areas of land. The process continues to this day, modern engineering techniques making possible such vast projects as the great barrage thrown across the mouth of the Ijsselmeer, which was completed in 1932.

Much of Holland lies below sea level, and in the early Middle Ages consisted of peat bog and salt marsh. The country lay at the mercy of the North Sea, which was held in check only by the sand dunes along the coast. From time to time the dreaded combination of Spring tides and northerly gales would raise the level of the sea far above its normal height, and devastating floods would ensue. The land painfully reclaimed through years of labour was lost again, and the effects of natural drainage were nullified. Such an inundation, occurring at the beginning of the fourteenth century, resulted in the formation of the Zuider Zee. Early flood defences were planned on a purely local basis to protect individual cities and districts. In so doing they sometimes placed other regions in peril, and it is probable that reclamation work around the mouth of the River Maas contributed to the Great Flood of November 1421, which caused enormous loss of life. Understandably therefore, it was in Holland that systematic and comprehensive reclamation schemes were first put in hand, and Dutch engineers soon became the acknowledged experts in work of this sort.

The methods of raising water employed in medieval Europe were of very limited use in the drainage of great tracts of flooded land. To make headway in these conditions a machine was required that would work continuously, day after day without respite. The sluggish rivers of a flat country could seldom supply the power required to drive the wheel of pots, and in any case this device was more suited to irrigation than drainage. In the absence of swiftly flowing streams, the only alternative source of power was the wind, and it is probable that the first windmills adapted for drainage work appeared in Holland during the early years of the fifteenth century. Rivers were embanked, and mills were set up along the dikes to lift water from the polders below by means of scoop wheels and Archimedean screws. Their numbers increased quickly,

De Peilmolen, *a wip mill at Groot Ammers—one of a series of large mills set up in the seventeenth century to drain the polder lying to the south of the River Lek*

and the windmill soon became established as a familiar feature of the Dutch landscape.

In the medieval post mill stones and machinery were mounted in the buck, which was turned bodily to face the wind. In drainage work a complication arose, since the position of the scoop wheel was necessarily dictated by the trough of timbering or brickwork in which it was to turn. The problem of power transmission from a rotating cap to a stationary base was solved by the introduction of the hollow post mill. Known in Holland as 'wip' mills, these curious structures, with their gaily painted caps perched above steep pitched roofs of thatch, are at once the most distinctive and picturesque of all Dutch mills.

The basic principle of the hollow post mill is explained by the name. The main post, on which the cap turned, was built up in the form of a hollow column through which the driving shaft passed down into the base of the mill. The cap was much smaller than the buck of the traditional post mill, since it housed only the great brake wheel and wallower. The latter wheel, following ancient practice, invariably took the form of a lantern pinion of timber. The base was developed into a square enclosure with low walls of brickwork, and a pitched roof which followed the line of the raking quarter bars. One half of this space was occupied by the pit wheel, driven by a second pinion at the foot of the main shaft. The other half, lit by dormer windows set in the thatch, provided cramped but snug accommodation for the miller and his family. In some mills the scoop wheel itself was housed within the base, in others it was fitted externally. It consisted of a narrow wooden paddle wheel, with long backward sloping blades, designed to fit closely between flank walls of brickwork or boarding. As it revolved, the water from the polder was driven upwards against a curved breast, to discharge at a higher level through a hinged wooden sluice. Like a conventional water wheel, the scoop wheel was particularly subject to decay, and while early examples were constructed of timber, iron was later introduced for the central frame, and finally for the whole wheel. The angle of the blades was arranged to assist in throwing off the water at the top of the effective lift, which amounted to rather less than one quarter of the diameter of the wheel. The sluice was hinged to close under back pressure, like the gate of a lock, and so prevent water running down into the polder when the mill was standing idle.

LEFT
Wip mill—the bearing at the top of the hollow post

ABOVE
Worn staves in the wallower of a large wip mill

LEFT
The windshaft and brake wheel

BELOW
Sails cast shadows across the windows of the miller's house, formed in the underframe of the mill

LEFT
The hinged iron sluice gate at Herringfleet drainage mill, Suffolk: note the wooden blades of the scoop wheel

FAR LEFT
The iron scoop wheel of a large wip mill

RIGHT
A thatched polder mill dating from 1740 on the Kinderdijk, near Rotterdam

BOTTOM
Cars drive under the lofty ladder and tail pole of a large Dutch wip mill. The church of Groot Ammers, in the background, stands on the dyke embanking the River Lek, many feet above the low lying polder

BELOW
A small Dutch meadow mill

Wip mills varied considerably in size, and hence in the living accommodation they could provide. Very often this was supplemented by the erection of a low cabin or 'summer-house' close by. The size of the mill also dictated the type of winding gear used, large examples being turned by winches mounted at the foot of the ladder, while smaller mills employed the simple tail pole. The smallest of all were fitted with wooden vanes, turning like weathercocks to face the wind, and many of these remain in use driving simple centrifugal pumps of wood. An automatic 'quartering' device is incorporated which feathers the sail assembly when the force of the wind becomes excessive.

The latter part of the sixteenth century saw the emergence of a new type of drainage mill. This was the great polder mill, a huge timber smock mill clad in thatch, the structural form of which has been briefly described in a previous chapter. Larger and more powerful than its predecessors, the polder mill was employed on the more ambitious reclamation schemes then being undertaken by the Dutch engineers. Practical considerations limited the size of scoop wheels, and thus the height through which water could be raised in a single lift. But this obstacle was overcome by the erection of 'gangs' of mills, three, or even four, working together to lift water in stages from the low lying polders. In this way quite extraordinary results were achieved, and lifts of 15 feet were not uncommon. Alternatively, by using an Archimedean screw in place of a scoop wheel, the Dutch millwright could achieve comparable lifts with a single mill, and this ancient device was particularly favoured in North Holland and Friesland.

The Archimedean screw consisted of a continuous spiral chamber formed around an inclined central shaft. By rotating the shaft, water was drawn up through the spiral passage and discharged from its upper end. Manually operated screws have been used in the East for irrigation since classical times, a surprising survival among backward farming communities of a relatively sophisticated contrivance. Screws driven by water wheels were sketched by Leonardo da Vinci, but it is probable that natural power was first effectively harnessed to this device by the millwrights of Holland.

Tail pole and anchor posts at the mill of the Marne polder, near Bolsward, Friesland

ABOVE
An Archimedean screw of steel

ABOVE LEFT
An Archimedean screw removed from a Friesian spider mill

LEFT
A view looking up into the thatched cap of a Friesian polder mill. The wallower has been repaired with steel angles: the brake wheel and brake are visible on the left

LEFT
The gearing between main shaft and water screw (right) in the Marne polder mill

OPPOSITE
The date board and sails of a Kinderdijk mill

BELOW
A brick polder mill dating from 1738, on the Kinderdijk

The screws employed in the great polder mills were constructed, like scoop wheels, at first of timber and later of metal. Wooden screws were beautiful and complex objects, requiring a high degree of craftsmanship on the part of the millwright. They were mounted in closely fitting troughs of brickwork, and the water drawn up from the polder was delivered through an automatic sluice to a culvert formed beneath the floor of the mill. Power was transmitted by bevel gearing, from a vertical drive shaft extending through the full height of the mill.

Both smock and tower mills were employed for drainage work. The largest, like the magnificent examples which survive from the early eighteenth century at Kinderdijk near Rotterdam, afforded ample accommodation for the miller and his family. Machinery was confined to the cap and to the ground floor, where the huge pit and scoop wheels were located on one side of the entrance passage. On the other side, behind a wooden partition, was the living room, complete with fireplace and painted panelling. The upper floors were divided into bedrooms with built in bunks, in which the miller's children no doubt learned to sleep, like mice in a clock, through the noise and vibration of the working mill.

Various smaller types of drainage mill were also to be seen in Holland. Most handsome of these was the Friesian 'spider', a miniature wip mill, with boarded cap and pantiled roof below. These little mills, which were fitted with Archimedean screws, were not large enough to provide living accommodation. The miller would set the sails and turn the mill into the wind, then leave it to work away untended until a change in the wind made another visit necessary.

The most simple drainage mill of all was the 'tjasker', another Friesian type, consisting of little more than an inclined windshaft carrying an enclosed Archimedean screw. The upper end of the shaft terminated in a pair of diminutive sails, and the whole structure, being mounted on trolley wheels, could be trundled round to face the wind. In these little field mills the wind driven screw was reduced to its barest essentials, and few examples survive today.

Following the lead of the Netherlands, other European countries put reclamation schemes in hand during the sixteenth and seventeenth centuries. Since the Middle Ages, efforts had been made to protect low-

1738

LEFT
Kinderdijk—three of the thatched smock mills of the 'Overwaard'

BELOW LEFT
A built-in bunk in the living room of a polder mill

BELOW RIGHT
Kinderdijk—a doorway in the base of the tower gives access to the great iron scoop wheel

LEFT
Kinderdijk—the brake wheel (left) and wallower of a brick polder mill, viewed from below

BELOW
A spider mill near Bolsward, Friesland

lying land from flooding by the erection of sea walls and embankments. Such local schemes were small in scale and met with limited success, and it was only with the adoption of the windmill for raising water that it became possible to envisage the actual drainage of large areas of marsh and fenland.

In Britain a drainage mill was at work near Holbeach in Lincolnshire in 1588, and the windmill soon became as familiar in the flat terrain of Eastern Britain as in the polders of Holland. The systematic drainage of the fens was at first undertaken by groups of 'adventurers' who, as speculators, financed the work in return for an agreed acreage of the reclaimed land. They met with determined opposition from the local inhabitants, the 'slodgers', who saw their traditional means of livelihood in danger. Living a curious, almost amphibious, existence as wild fowlers and fishermen among the reed beds and pools of the remote fens, these people cherished a tradition of fierce independence. By breaking down banks and wrecking sluices they defeated many of the projects of the early 'drainers'. A principal target for their hostility was Cornelius Vermuyden, the brilliant Dutch engineer engaged to carry out the drainage of the Bedford level. This great undertaking, backed by the Earls of Bedford, was begun in 1634. Vermuyden was as determined as his opponents, who resented his overbearing manner and scorned his 'cranks and whirligigs'. In spite of all opposition he had achieved a large measure of success by 1655, when he finally relinquished his command of the project, and retired into obscurity.

The battle for the fens was still far from won, however. The main efforts of the 'adventurers' had been directed towards the embankment of rivers, the clearing of existing water-courses and the cutting of new drains to carry river water away to the sea. A host of windmills sprang up to drain the reclaimed land, but they were unequal to the task and seasonal flooding still affected wide areas. From time to time, particularly severe winters would bring disastrous floods, destroying dikes and sweeping away mills, and it was not until the nineteenth century that the 'drainers' finally gained the upper hand.

As drainage progressed a curious side effect became apparent. The ground began to sink, and this general subsidence of the peat was accelerated when steam engines were introduced during the first half of the nineteenth century. The slow-running, low pressure beam engine of the period was well suited to the steady continuous work of drainage, and better able than the mill to deal with the ever increasing lifts required. Above all, it was not subject to the vagaries of the wind, and from the middle of the century the number of windmills began to decline, many finding themselves left high and dry by the shrinking peat. The last survivor, a small wooden smock mill, stands in Wicken Fen, which has been preserved undrained as a nature reserve. Paradoxically it is now necessary to lift water into the Fen, and the little mill has been adapted to serve this purpose. She was built early in the present century by Hunts, a famous firm of millwrights from Soham, and originally stood in Adventurers Fen, some miles from her present home. Typical of smaller drainage mills, she is square on plan, with boarded sides and cap, and curved wooden casings on either side to enclose the projecting blades of the scoop wheel. The wheel itself is of timber, and turns in a trough faced with boarding and backed with puddled clay. The winding gear consists of a braced tail pole, anchored by a folding crutch. Shafts and gears are of iron, and the stocks of the common sails are bolted to an iron cross in the Lincolnshire manner. Restored and rebuilt, she stands as a lonely reminder of a sight once familiar throughout the fens. Fenland scenes painted during the early years of the nineteenth century show a countryside enlivened by scores of mills. Most

LEFT
A Friesian tjasker turning briskly in a fresh breeze. Water is discharging from the open end of the screw (right)

BELOW LEFT
Diagram of a drainage mill from The English Improver Improved *by Walter Blith, 1652*

RIGHT
Wicken Fen Mill—a detail showing the direct drive from main shaft to scoop wheel

BELOW
Wicken Fen drainage mill

were timber smock mills, and some attained a considerable size. In spite of Dutch influence, the large hollow post mill so typical of Holland never became popular in England. The principle was however adopted for small meadow mills, many of which consisted of little more than sails and tail vane, mounted on a stout hollow post.

The fenland mills have gone, and the tall chimneys of the pumping engines have followed them into oblivion. But a number of drainage mills may still be seen in the neighbouring county of Norfolk, where many remained in service until the Second World War. They were built to lift flood water from low-lying marshes along the coast, and to drain the inland pastures through which the slow Broadland rivers wind out towards the sea. Here, as in parts of Holland, the position had been complicated by the cutting of peat for fuel. This ancient practice of 'turbary', exercised by the marshmen throughout the Middle Ages, resulted in the formation of the Broads, a huge network of lakes and channels whose artificial origin has only been established in recent years.

In Norfolk, most of the surviving mills date from the nineteenth century. But the earliest example, a ruined tower on the River Bure near Acle, bears the date 1750, while at the other end of the scale the little mill at St. Olave's was built as recently as 1910. In many cases the work of drainage was seasonal, the mills being 'set down' during the summer months. The installation of steam engines was not economical in these circumstances, and this no doubt accounted for the survival of the mills. Now most have fallen into decay, and sheds of corrugated iron house the electric pumps which have taken their place. The marshman and his family often lived in the mill; for them it was a lonely existence, their homes standing in remote corners of the marsh, far from the nearest road. Not infrequently winter gales would breach the sea walls, flooding the marshes behind the dunes, and completely isolating the mills. Old marshmen recall that they would then make a brave sight, their scoop wheels churning the salt water into a smother of spume.

The typical Norfolk mill of late years consisted of a stout circular tower of tarred brickwork, surmounted by a handsome boat-shaped cap of boarding, with galleries and fantail. A semicircular wooden

RIGHT
The drainage mill at Horsey Mere, Norfolk: the present mill was built in 1912 to replace an earlier structure. To reduce wind resistance, the sails have been restored without shutters or striking gear

FAR RIGHT
The drive from main shaft (right) to the turbine pump at Horsey Mere

BELOW
Thurne Mill, Norfolk—the gallery around the cap was a common feature of Norfolk mills

BELOW RIGHT
Horsey Mere—the massive iron brake wheel and wallower

Herringfleet Mill, Suffolk, mid nineteenth century. The semicircular wooden hoodway (right) encloses the scoop wheel

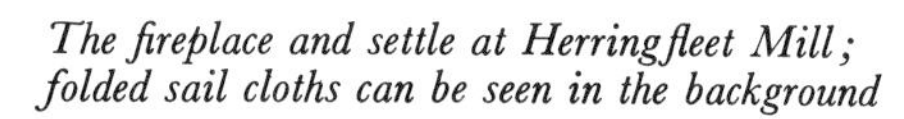

The fireplace and settle at Herringfleet Mill; folded sail cloths can be seen in the background

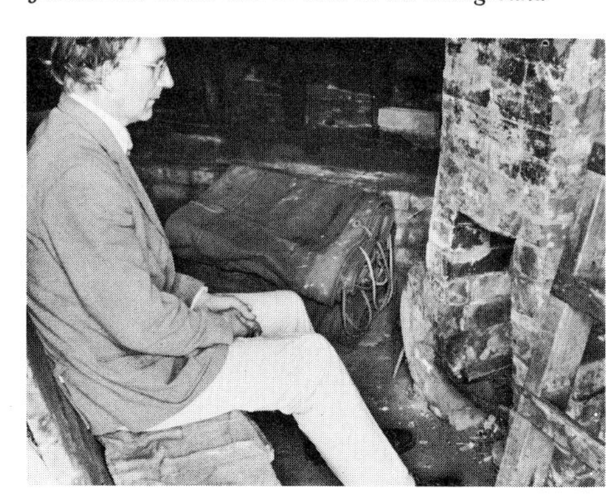

Herringfleet Mill—the tail pole braced in the Dutch manner

Herringfleet Drainage Mill, Suffolk

The drawing shows the structure and machinery of a small English drainage mill of the smock type, with common sails and braced tail pole. Brake wheel, wallower, and main shaft are all of timber, the windshaft and pit wheel of iron. The scoop wheel, 16 ft. in diameter and 9 in. wide, consists of a central iron frame carrying wooden blades, and is enclosed by a semicircular hoodway of boarding, which has been omitted from the drawing. Scale in metres

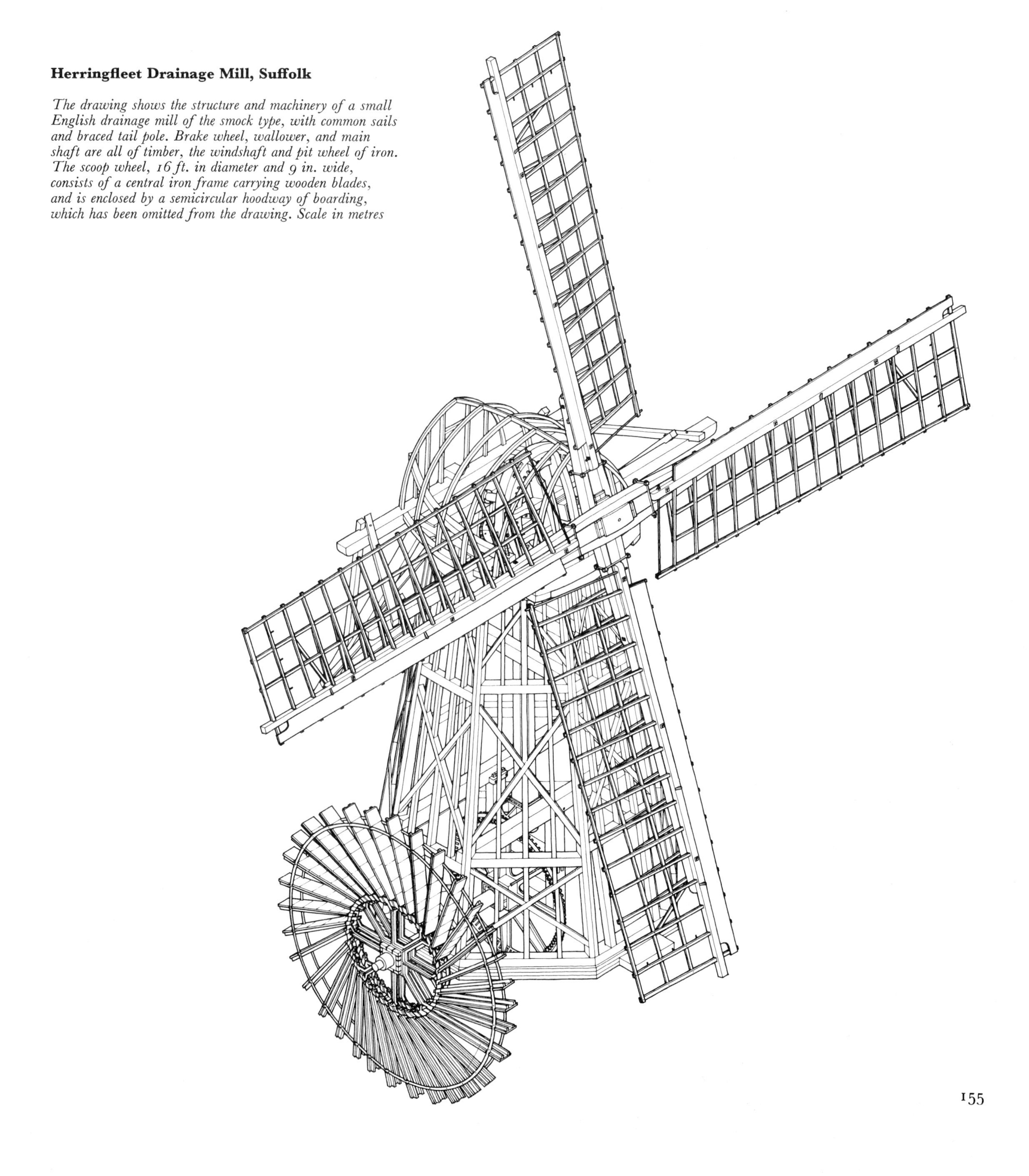

casing, or 'hoodway', against the wall of the mill enclosed the scoop wheel, which was driven through bevel gearing from the main vertical shaft at a speed of about 100 revolutions per minute. Occasionally pumps driven by a crank shaft were fitted instead of a scoop wheel but this arrangement was unusual. A more common alternative was the turbine pump, developed in the mid nineteenth century and claimed to be a great deal more efficient, in a steady breeze, than the old fashioned wheel. The turbine was employed in a number of the later mills, including the fine brick tower mill on Horsey Mere, now happily preserved. The high mill at Berney Arms on the River Yare, most impressive of all the Norfolk drainage mills, is also maintained in perfect condition. Standing over 70 feet in height, this mill is unusual in several respects, having been built to serve a dual purpose. A pair of edge runner stones ground clinker for cement, while the primary function of drainage was accomplished by a large free-standing scoop wheel, located some distance from the base of the tower.

It is probable that a number of these sturdy brick towers replaced earlier timber structures, like the smock mill which survives beside the River Waveney at Herringfleet in Suffolk. Too small to house a family, this mill still contains a fireplace on the ground floor and a rough wooden settle where the marshman could sit to dry his clothes on a rough winter night. As in Holland, the smoke from the fire was discharged into the top of the mill, finding its way out through the cap, and in the course of years it imparted a fine dark patina to the timber. Herringfleet Mill was built in the mid nineteenth century, and is still in working order. With common sails and braced tail pole, this little building bears a strong resemblance to the mills of Holland, and stands as a memorial to the engineers from across the narrow seas who introduced their techniques of land drainage into Eastern England during the seventeenth century.

Wind pumps with 'jib' sails in the Plain of Lassithi, Crete

Wind pumps near Famagusta, Cyprus

The use of windmills for drainage has largely died out with the introduction of alternative and more reliable sources of power. But in many countries the pumping mill still raises water for irrigation and supplies the needs of isolated farms. Tracts of land, formerly too dry for farming, have been rendered fertile by the tapping of natural reservoirs of water below the ground and, in the absence of electricity, wind power has provided the means of bringing this life-giving water to the surface. Conventional windmills have been used for irrigation, particularly in Portugal, where the wheel of pots, driven by a mill, is employed for shallow lifts. But the machine which has become associated above all others with this type of work is the American wind pump, still a familiar sight throughout the world. This uncompromisingly functional structure consists of a lightly built pylon of lattice steel work supporting an annular sail or wind wheel. A broad tail vane holds the mill in the eye of the wind, and at the same time blazons abroad the name of the manufacturer. A cranked windshaft imparts a reciprocating motion to the long steel pump rod which passes down through the pylon to the bore hole beneath. The type of wind wheel in most general use consists of a circle of curved metal vanes, arranged like the petals of a flower. The illustration shows a forest of these pumps located near Famagusta in Cyprus; the numbers are exceptional, but the same pumps might be found on the chalk downs of England or in the Australian outback. A more local type of wind wheel favoured in Majorca is polygonal in shape, and is fitted with sets of narrow vanes which can be opened or closed like the sticks of a fan. Most distinctive of all are the wind pumps of Crete, which are powered in the manner of the eastern Mediterranean, by triangular sails of white cloth spread on wooden spars. These are to be found in great numbers in the plain of Lassithi, where a multitude of mills, moving in the breeze like a field of white campions, produces what must surely be one of the strangest man-made landscapes in the world.

The pumping mills once to be seen along the English coastline were not all employed in the drainage of marshland. The production of salt by the evaporation of sea water was of prime importance for many centuries, when English herring busses supplied huge quantities of salted fish to the Roman Catholic countries of Europe. The sea water was admitted to wide shallow pools, and allowed to evaporate in stages. As the salinity of the water increased it was raised to higher basins, and small windmills were commonly employed for this work. In the final stage of the process the concentrated brine was poured into iron pans and heated over fires of wood or peat. Salt is still produced by natural evaporation in warmer lands, but in Britain the unequal struggle with the northern climate has long since been abandoned, and only the occasional embankments and channels of old salterns along the coast bear witness to what was once a flourishing industry. Old prints of Lymington in Hampshire show scores of small skeleton windmills standing in the salterns, but none of these survive. They appear to have been hollow post mills driving scoop wheels or pumps, and comparable with the simple Norfolk meadow mills so carefully drawn by Constable.

Prominent windmills were familiar landmarks to the seaman, helping him through the hazards of coastwise navigation along the shores of the North Sea. They were even used at sea, providing assistance of a more positive kind to ships engaged in the Baltic timber trade. Bulky deck cargoes tended to prevent access to the pumps, and many Scandinavian barques of the nineteenth century were equipped with tall lattice windpumps, driven by canvas sails. Known as 'onkers', from the melancholy clanking of their pumps, these vessels provided one of the more bizarre instances of the employment of the windmill in the service of man.

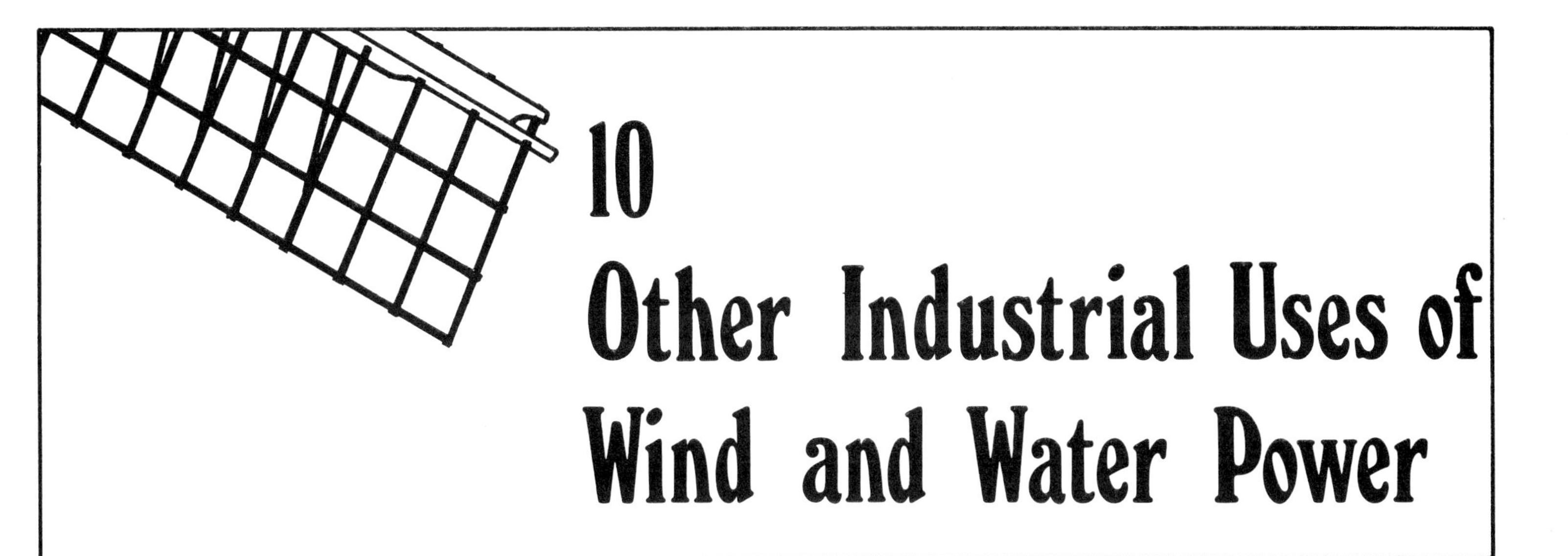

10 Other Industrial Uses of Wind and Water Power

The Cat, *a dye mill on the Zaan*

Among people unaccustomed to the most simple machinery, it required an immense effort to adapt the turning wheel to produce the reciprocating motion required for so many processes performed by hand. It seems likely that this step was first taken in China, where water driven bellows are recorded as early as AD 31. Used to produce the draught required for smelting iron, they were actuated by a cord attached to an eccentric peg set in the face of a wooden wheel. A simple horizontal water wheel mounted on the lower end of the shaft provided the motive power. It is strange that the principle of the crank, employed in this early machine, was to lie dormant for so many centuries.

The first industrial mills to appear in the West were probably used for fulling cloth. But by 1010 water power was assisting in the production of iron in Germany, and by the time of the Domesday survey the rents of two Somerset mills were being assessed in blooms of iron.

Early methods of smelting ore did not achieve high enough temperatures to produce molten metal. Iron was deposited on the stone base of the hearth in the form of a viscous sponge-like mass or bloom, which was consolidated and scoured by hammering. The violent pounding action of the smith was reproduced mechanically by the use of a heavy pivoting hammer, its haft depressed and released by cams projecting from the axle of a water wheel. A similar machine had been used for hulling rice in China since the second century, but it was not until the early Middle Ages that the sound of the tilt hammer was to become familiar in the forges of Europe.

With the introduction of the blast furnace in the fifteenth century, served by water powered bellows, it became possible to reduce large quantities of iron to the molten state, and to cast such objects as cannon. But the brittle, intractable pig iron produced remained unsuitable for many of the requirements of the smith, and the tilt hammer was again employed to produce a more malleable material. This was achieved by a process of re-heating and hammering, which removed impurities and reduced the carbon content of the metal.

Tilt hammers also performed other functions. Until the advent of the rolling mill they were employed to produce sheet iron from the bar, and the subtle shaping of weapons, scythes, mattocks and shovels,

ABOVE
A Surrey hammer pond

ABOVE LEFT
A water powered rice mill in Sumatra

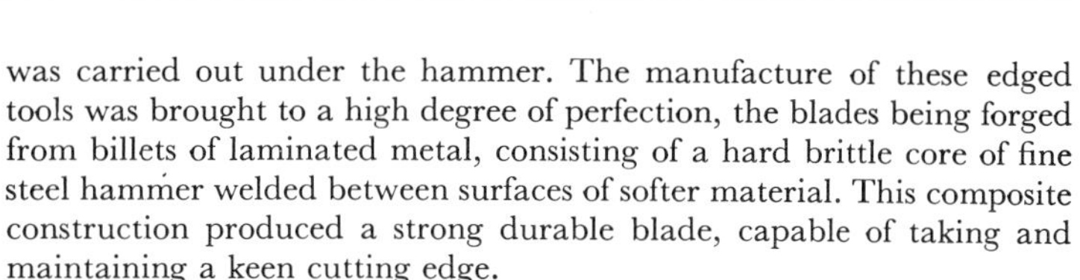

was carried out under the hammer. The manufacture of these edged tools was brought to a high degree of perfection, the blades being forged from billets of laminated metal, consisting of a hard brittle core of fine steel hammer welded between surfaces of softer material. This composite construction produced a strong durable blade, capable of taking and maintaining a keen cutting edge.

The earliest tilt hammers were of simple design. Studs projecting from the axle of the water wheel engaged directly with the haft of the hammer, which was pivoted between two timber uprights. In more developed examples, the axle carried a separate trip wheel within the mill, with cams projecting from its rim. The speed of the hammers varied according to the process for which they were to be used. One hundred and fifty strokes per minute was reckoned to be a suitable rate for the drawing of blades, a considerably faster rate being employed for the forging of steel billets. In order to achieve these rapid speeds, it was necessary to restrict the natural swing of the hammer. This was done by mounting a recoil spring above the haft, in the form of a long resilient plank of timber, to cushion and return the rising hammer head. Alternatively, a stout iron-bound recoil block was set in the floor of the forge, below the tail end of the haft, to perform the same function.

The fuel used for smelting iron was charcoal, and huge tracts of forest land were cleared of trees to feed the furnaces of medieval Europe. The widespread felling of trees eventually threatened the supply of timber for ship building, and in Britain a law was passed in 1585 designed to prevent iron masters from gathering fuel outside the boundaries of their own estates. In the weald of Kent and Sussex, once densely wooded, the transformed landscape of rolling farmland survives as the most permanent legacy of the vanished iron industry. More obvious to the casual observer are the great mill ponds, where streams have been dammed to drive the hammers of mills long since fallen into decay. Place names often provide evidence of former industrial activity, the word 'hammer' occurring frequently in Sussex, in the names of farms and woods, mill ponds and streams. Many of the craftsmen who worked the forges were foreigners, or skilled men brought in from other industrial areas. They tended to form their own communities, having little in common with the local countrymen, to whom

Tilt hammers at Finch's Foundry, Sticklepath

The hammers are mounted in a massive frame of timber, and swing on trunnions between iron cheeks. They are of 'tail helve' type, their butts depressed by cams projecting from iron tappet wheels. One wheel has 16 cams, the other 12, producing different rates of striking. The main axle is of timber and carries a primary gear wheel which supplies power, through lay shafts, to other machinery. A wheel mounted on the inner (left hand) end of the axle drives the shears (left foreground), by means of an ingenious hooked strap. The pitch-back wheel, of composite construction, is of 12 ft. 4 in. diameter, and is served by an overhead launder of timber, not shown in the drawing. Scale in metres

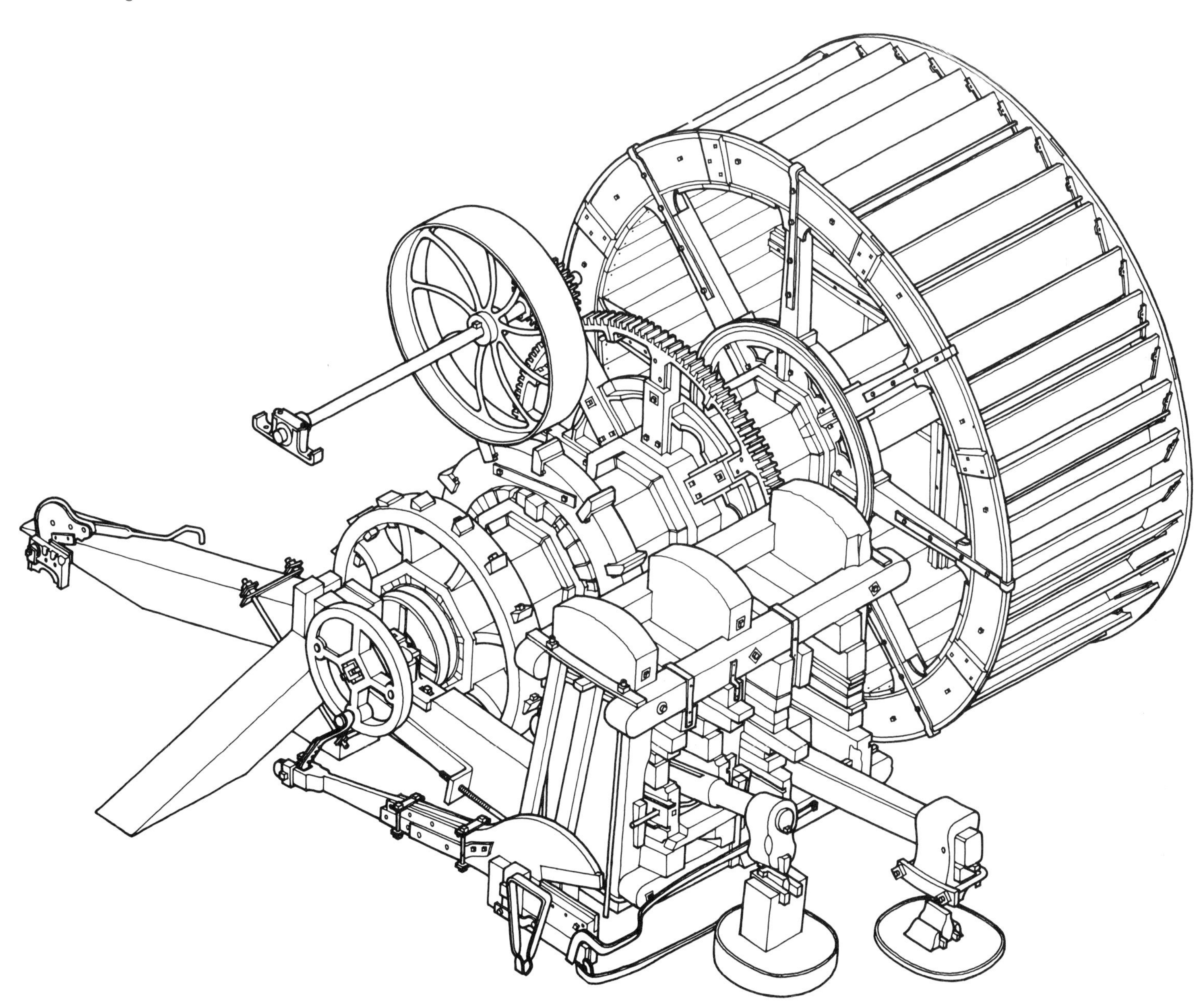

LEFT
Water power employed for drawing wire, from Pirotechnia *by Vanocci Biringuccio, 1559*

BELOW LEFT
A water driven fan in use for mine ventilation early in the sixteenth century, Agricola

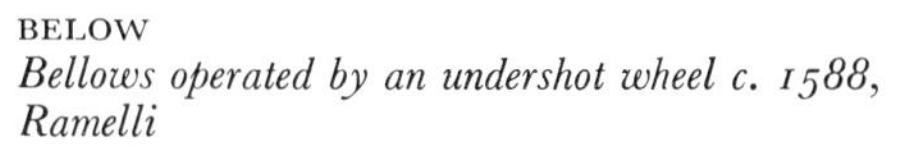
BELOW
Bellows operated by an undershot wheel c. 1588, Ramelli

LEFT
The tilt hammers at Finch's Foundry, Sticklepath, Devon, before restoration

BELOW
Finch's Foundry—when not in use, the hammers were held out of gear by stout wooden struts slipped into place under the shafts. The anvils are here shown capped, to prevent accidental damage

they were 'strangers at the hammer', a phrase recurring in the registers of Wealden churches. It is perhaps for this reason that few local traditions of the industry survive, even in villages which must once have rung with the noise from the forges. This was considerable; at Boldre in Hampshire the inhabitants could hear the thud of the great hammers at Sowley pond, well over three miles away. When the sound carried clearly, it was said to foretell rain.

In addition to driving hammers, water wheels supplied power for a number of other processes connected with the production of metals. They worked bellows, and later fans, to serve hearths and furnaces; and by the middle of the sixteenth century, water power was in use for drawing wire. Mills for slitting and rolling bar iron followed, and by the eighteenth century complicated water powered machinery was being employed for the boring of gun barrels.

One of the last water powered forges to work in Britain was the Finch Brothers' foundry at Sticklepath in Devon. Behind the village the great bulk of Cawsand Beacon marks the northern boundary of Dartmoor; and the little River Taw, running down from the high moor; once provided power for several mills. The foundry was established here, beside the village street, in 1814, on the site of earlier corn and woollen mills, and it ceased production in 1951, following the collapse of the roof. To the end of its working life it enjoyed a high local reputation in the West Country for the quality of its products. The principal machinery survives, and much of it has been expertly restored to working order. It clearly illustrates three different applications of water power, separate wheels driving the tilt hammers, the grindstone, and the fan. The machinery of the forge was subjected to far more violent stresses than that of the corn mill. The rapid striking action of the hammers produced a shuddering vibration that shook loose wedges and made constant maintenance and repair essential. Thus at Sticklepath the great timber axle, which measures over two feet in diameter, was renewed before the closure of the foundry) in African hardwood, a suitable baulk of British timber being no longer obtainable. The hammer shafts themselves, once of timber, have been renewed in steel at some time in the past. When restoration began the wooden shafts were still remembered in the village, an old man recalling how he had once, as a

ABOVE
Finch's Foundry—the shears

TOP
Water powered tilt hammers in use in a Sheffield forge, 1933

RIGHT
Finch's Foundry—the grindstone. The craftsman lay on the adjustable 'stretcher' (top left); a worn stone lies in the foreground

FAR RIGHT
Finch's Foundry—the fan

BELOW
Finch's Foundry—a drop hammer operated by friction drive from an overhead pulley

boy, been employed to keep them wet so that the hammer heads would not work loose. The shafts swing on trunnions mounted between heavy iron cheeks, which are in turn supported in a massive frame of timber, and held in position by a motley assortment of timber packs and wedges. A great pair of shears is mounted beside the hammer frame, operated by a connecting rod linked to a wheel on the end of the main axle. Each tool was roughly shaped by the hammers, trimmed with the shears, and finished by hand on a separate anvil. Blades were sharpened in the grinding house, which formed a separate part of the mill complex with its own overshot wheel supplying power to the stone. The craftsman lay face downwards on a curious adjustable platform above the grindstone, holding the blade at arm's length below him. At the downstream extremity of the foundry stands the fan house. Here a four bladed iron fan, driven by a third water wheel, supplied a forced draught to the various hearths within the main building, the air passing through a system of ducts laid below floor level.

For centuries weapons and tools, cooking pots and castings, were produced in small forges and foundries like the example which survives at Sticklepath. Prosperous iron masters such as Richard Lenard, whose likeness can still be seen in Lewes Museum, staring proudly out from one of his own firebacks of Sussex iron, were men of considerable local standing. Their fine stone houses command attention today, but the foundries themselves were comparatively humble buildings which have either disappeared or have been converted to other uses.

The production of iron and steel implements required the skills of various different crafts. In the old established industrial centres, these tended to maintain their independence, and small mills were often equipped to perform a single operation. Typical of these were the grinding mills, set up wherever falls permitted on the streams around Sheffield. An example survives in Shepherd Wheel on the River Porter, a small gable ended building of stone, set in the Whitely Woods. Here blades were being ground in the late sixteenth century, and the mill remained in commercial use until the 1930s. When the whole manufacturing process was carried out by a single firm, a complex of buildings would result. An industrial unit of this sort, dating from the eighteenth century, can be seen at the Abbeydale Works, Sheffield, where the forges, foundry, tilt shop and grinding shop are ranged around a courtyard, while the warehouse, manager's house, and workmen's cottages complete the group.

Mention has been made of the sack hoist, the first auxiliary machine to be introduced in the corn mill. In mining districts very much larger hoists, powered by water wheels, were employed as winding engines, raising the ore in buckets from the bottom of the shaft. In earlier days this work had been performed, slowly and laboriously, by manually operated winches, an arrangement which survives in the traditional winding gear to be seen at well heads. By the middle of the sixteenth century, however, both the horse whim and the water wheel were in use, and water powered hoists were illustrated by Agricola. They had no provision for throwing the hoist barrel out of gear, and it was therefore necessary to stop the wheel and reverse the direction of rotation between lifts. It is interesting to note that a remarkably similar device, powered by opposing turbines, was being installed in German mines no less than three centuries later.

The power driven windlass was not confined to the mine, and water wheels were in use, possibly as early as the twelfth century, on canals and navigable rivers in the Low Countries. Here they were employed, before the introduction of pound locks, to work barges upstream against the surges of water sent down by the stanches of primitive flash locks. Similar winding gear, both wind and water powered, was also used in China during the medieval period to tow boats upstream against a swiftly flowing current. During the great age of canal building, considerable changes of level were frequently negotiated by long ramps, or inclined planes. Wagons, or 'tub boats' mounted on trolleys, would be drawn up the incline by a winding engine, and where conditions allowed, this would be powered by a water wheel. On the Tavistock canal, opened in 1817, wagons loaded with copper ore were lowered down such an incline to Morwellam Quay on the River Tamar, a drop of 237 feet. The cable drum was in this case driven by a large overshot wheel, and the weight of ascending wagons was used to counterbalance those on the way down. A tripping device was incorporated to put the drum out of gear as wagons reached the top of the slope. Similar inclined planes were sometimes used in mines, and a notable example, powered by a water wheel 40 feet in diameter, was constructed to serve the Wheal Friendship mine at Mary Tavy during the early years of the nineteenth century.

A number of industrial processes made use of the edge runner mill, a device suited to crushing rather than grinding, and described in some detail in Chapter 1. Early mills of this type were worked by animals or slaves, and it was only in the seventeenth century that the water powered edge runner mill appeared in Europe. In China, however, it had already been in use for many centuries, having been adapted to water power by about AD 400. The edge runner mill was used in the preparation of raw materials for a very wide variety of products including spices and snuff, pipe clay, china and cement. It was also used in the manufacture of gunpowder, and a remarkably complete installation dating from the late eighteenth and early nineteenth centuries survives among the ruins of the Royal gunpowder factory at Faversham in Kent.

This small mill at Faversham was one of a group of four, whose foundations may still be traced. Power was supplied by two iron breast-shot wheels, one of which remains *in situ*, and the primary gearing resembles that of the conventional watermill. The mill could be stopped by lifting the spur wheel out of gear, an operation performed by turning a threaded collar on the vertical shaft by means of hand spikes. The

RIGHT
A water driven edge runner mill, fitted with a single stone, from Novo Teatro di Machine et Edificii *by Vittorio Zonca, 1607*

FAR RIGHT
The Householder, *a mustard mill on the Zaan*

BELOW RIGHT
The Householder *getting under way. In common with others on the Zaan, this little mill was moved to her present site in recent years and is still at work*

BELOW
'*Richard Lenard, Founder, at Brede Fournis, 1636*'*—a Sussex Ironmaster*

PESTRINO CHE PESTA LA VALONIA

Gunpowder Mill, Faversham, Kent

The edge runner mill illustrated formed a part of the Chart Powder Mills at St. Anns, Faversham. The breast-shot iron wheel, 16 ft. 2 in. in diameter, originally served a pair of mills. Of the machinery of the second, which would have occupied the left hand background of the drawing, only the pit wheel survives, and this has been omitted. The primary drive was transmitted through a bevelled wallower, to the vertical iron shaft, and thence, by spur gearing, to a second shaft of timber, around which the runner stones revolved. This timber shaft and spur wheel remain, but the edge runner stones and their supporting yoke do not, and are shown restored in the drawing. Also restored are the wooden starts and sheet iron floats of the water wheel. Scale in metres

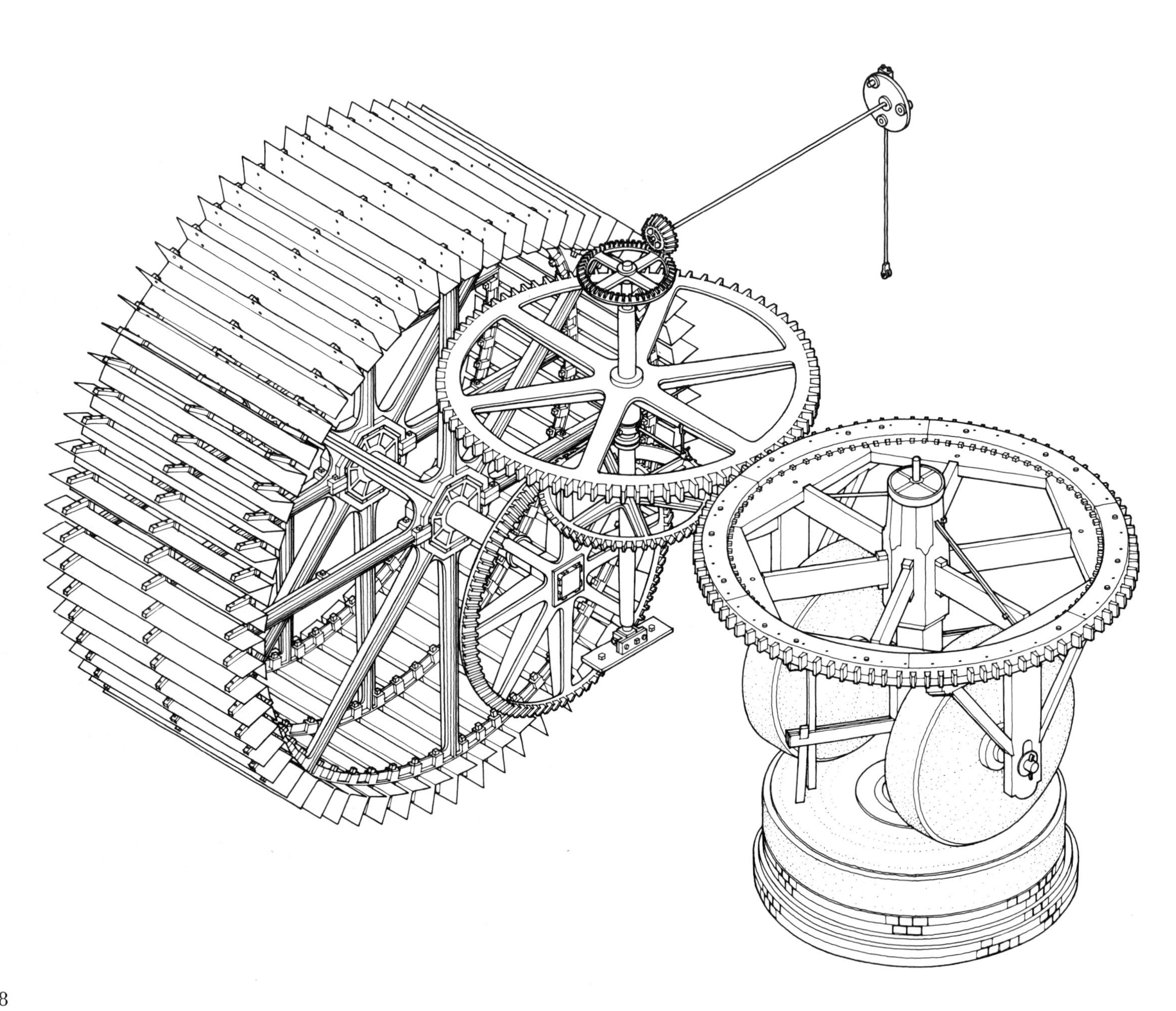

The gunpowder mill at Faversham—a view of the machinery in the course of restoration, showing the great spur wheel (top), the wallower and part of the pit wheel

The gunpowder mill at Faversham, before restoration. The central water wheel drove two pairs of edge runner stones: the circular base for the nearer pair is visible (bottom left)

A Zaan oil mill, Het Pink, *built in 1620 and restored in 1939*

The Seeker—*a set of iron-shod stamps poised above the pressing block: the edge runner stones occupy the foreground, left*

stones themselves were missing, but have been replaced by a pair from another site as part of an admirable restoration scheme. An unique feature of the powder mill is recalled by the iron connecting rod, driven through bevel gearing from the main shaft. It is probable that this once powered a small pump, which raised water to a tank in the roof space, as one of many precautions against fire. Explosions were an ever present hazard, and powder mills were designed and constructed to prevent, as far as possible, the chance of accidents occurring, as well as to minimise the resulting damage. Thus at Faversham, with the exception of one massive gable wall intended to withstand blast, the building itself was lightly constructed of weather-boarding on a timber frame. When in use, powder mills were kept scrupulously clean and free from dust. Machinery was liberally oiled and greased, and above the stones themselves, leather flaps were nailed over exposed iron bolts and nuts to avoid the accidental striking of sparks. But in spite of all precautions, here as elsewhere, disastrous explosions did occur from time to time, and the belts of tall trees, which now form an idyllic setting for the ruins, were originally planted to localise damage.

Edge runner stones may still be seen in use on the Zaan in Holland, although here wind and not water provides the motive power. A number of oil mills survive in the area, and one of these, named *The Seeker,* remains in active service, producing oil from groundnuts. Built in the early seventeenth century as a drainage mill, *The Seeker* was converted to her present use in 1671. As in other industrial windmills, her great thatched tower opens, at ground floor level, into a long gabled building of timber, with walls of painted weather-boarding. The interior is dominated by the ponderous runner stones, painted red and blue, which swing rumbling and creaking around their circular track near the centre of the mill. The flat bed stone is mounted on a brick plinth, and the nuts to be ground are spread across its surface in the path of the circling stones. After grinding, the meal is raked off, and spread in shallow pans to heat over a brick hearth. An iron paddle, driven by gearing from the main shaft above, slowly stirs the roasting meal, while the smoke from the pans drifts up into the tower of the mill overhead.

Another ancient machine may be seen in use in *The Seeker,* completing the process of extraction. This is the stamp mill, which employs heavy

LEFT
A stamp mill in use for crushing ore in Saxony, early in the sixteenth century, Agricola

RIGHT
The Schoolmaster *at Westzaan, built in 1692: drying sheds can be seen in the foreground*

BELOW
Interior of The Seeker, *a working oil mill showing the yoke of edge runner stones*

baulks of timber to compress the heated meal. Power is supplied by a horizontal driving axle or camshaft on the floor above, equipped with projecting studs which engage with the heads of the rams and stamps, lifting them in turn and allowing them to fall back under their own weight. Beneath them lies the pressing block, formed from a roughly squared tree trunk, with cavities hollowed out from the solid timber under the feet of the stamps.

The meal is packed into porous bags, which are then placed in the block and compressed by means of folding wedges, driven home by blows from a square wooden ram, while the oil trickles down into shallow pans below. When the miller judges that sufficient pressure has been applied, a second ram is brought into use to release the wedges, and the meal is removed in the form of a consolidated cattle cake. If further extraction is required, the cake is broken up, packed into a further set of cavities in the pressing block, and pounded with stamps, the feet of which are shod with iron. Finally the meal is again bagged and pressed, to emerge once more as a rectangular block of cake.

Like the tilt hammer, to which it was related mechanically, the stamp mill was introduced during the Middle Ages, and was possibly first employed for fulling cloth. Stamps, or falling stocks, remained in use for this purpose until the present century, and provided an accepted alternative to driving stocks of the type preserved at St. Fagan's Castle. Water powered stamps were at work in mining districts by the fourteenth century. Set up on the surface close to the mine shaft, they were used to break up the ore, in preparation for sorting and panning. Their early use in Saxony is attested by Agricola, and very similar installations, characterised by cumbersome horizontal drive shafts studded with projecting cams, were to remain in use in Devon and Cornwall throughout the nineteenth century. Materials and methods of construction might change, but the machine itself was simple and efficient. 'Wet stamps' were sometimes employed, a stream of water from the tail race being directed over the ore to wash away unwanted material during the crushing process. Stamp mills, like tilt hammers, were notoriously noisy, and the rhythmic thudding of the iron shod stamps earned them, in Cornwall, the name of 'clash mylls'. Even the rams and stamps of the wind driven oil mill may be heard from a considerable distance, and within the mill the noise is deafening.

Water powered stamping machinery was used in other industrial processes, notably the manufacture of paper. Linen and cotton rags were pulped by a battery of long wooden hammers, the heads of which were lifted and dropped in turn by a horizontal camshaft. Although commonly referred to as 'stampers', the hammers of a paper mill were more usually driving stocks very similar to those used in fulling. In the late seventeenth century, however, a more efficient mechanical pulping device was introduced in Holland, where the rags were disintegrated between fixed and revolving blades. During the eighteenth century this machine, known as the 'Hollander beating engine' was generally adopted in Britain.

In Holland paper making, in common with other industries, was forced to rely on wind power, and one great paper mill, *The Schoolmaster*, still stands intact at Zaandam, the sole survivor of a vanished industry. The thatched tower of the mill surmounts an immense drying shed, clad in green painted weather-boarding and roofed with red tiles. Tiers of horizontal hinged shutters in the walls allowed the air to circulate freely among the racks of drying paper, and the remains of similar louvred outbuildings are often associated with rural mills once used for paper making. The industry was not centralised, and mills were established, or more often converted, in widely separated

localities. One vital pre-requisite was an ample supply of pure water, and this no doubt accounted for the remote sites sometimes chosen. The craft was introduced to Britain from the continent, the first mill being set up at the end of the fifteenth century. Continental techniques were generally more advanced than those employed in Britain, and the names of many of the families associated with the early history of the paper making industry in this country betray foreign origins. Perhaps the most famous of these emigrés was Henri Portal, a young Huguenot refugee, who was apprenticed in Southampton and later, in 1711, set up as a paper maker at Bere Mill, on the clear head waters of the River Test in Hampshire. Its working life over, Bere Mill still stands today as a private house, while at Laverstoke close by the modern Portal Mill produces the fine paper which has been used for British banknotes for well over two centuries.

In the Middle Ages, and for centuries after, the slow and laborious conversion of timber by the use of hand saws was an inevitable part of every building enterprise. The desirability of some form of mechanical saw was clearly recognised, but the technical problems involved presented a formidable challenge to the medieval engineer. A primitive water powered saw was evidently in use by about 1270, and is illustrated in the sketchbook of Villard de Honnecourt. It employed an unsupported blade, drawn downwards by cams projecting from the axle of a water wheel and returned by a flexible pole spring. Whether de Honnecourt's saw can ever have functioned efficiently is very doubtful. It is certain that centuries were to elapse before mechanical saws were used to any extent, and that such master works of the medieval carpenter as the angel roofs of East Anglia were framed up entirely from hand-sawn timber.

The basic design of the reciprocating frame saw had been evolved by the late fifteenth century, and an illustration of such a device appears in Francesco di Giorgio's treatise on machines. This shows a blade set in a rectangular frame, arranged between a pair of vertical guides, power being transmitted through a connecting rod and crank from the axle of a water wheel. The timber to be sawn is laid on a wheeled trolley

LEFT
De Gekroonde Poelenberg, *a restored paltrok mill on the Zaan. Logs were raised to the open sawing floor by the crane (right)*

BELOW
A paltrok mill, De Held Jozua, *at Zaandam—the winding gear and sawing floor*

and moved forward by a ratchet mechanism at each stroke of the saw.

The first wind driven sawmill was built in Holland in 1592, and the following century saw many mills, both wind and water powered, established in Northern Europe. Much timber was still hand-sawn, however, and when the first British sawmill was set up at Deptford it was destroyed by the sawyers, who considered the new method a threat to their ancient craft. Sawyers were traditionally, and not surprisingly, thirsty, and inns still bear the name 'The Windmill and Sawyer', in celebration of this event. But despite such opposition, the sawmill soon became established in Britain, the water powered version being known at first as the 'Norway engine'. The new machine was able to generate sufficient power to drive a gang saw, in which a set of parallel blades were tensioned between the horizontal members of a massive timber frame. The reciprocating motion was achieved by the use of a crank, coupled to a connecting rod and cross-head.

In the windmill power transmission again presented special problems, and the first sawmills were of the 'paltrok' type, square timber smock

BELOW
Paltrok mill, De Held Jozua, *at Zaandam*

LEFT
De Held Jozua—*the sawing floor, showing saw frames (centre), flanked by ratchet wheel and pawls (left) controlling the timber feed*

BELOW LEFT
De Held Jozua—*an interior view, looking up. The crankshaft and driving pinion are visible (top left) and the timber connecting rod, in the breast of the mill (right)*

RIGHT
The Herdsman—*the drive to the crank shaft*

BELOW RIGHT
The Herdsman *at Leiden—a large nineteenth century sawmill of the smock mill type*

mills with fixed caps. These extraordinary buildings incorporated an open sawing floor, cantilevered out on either side of the body and covered by pitched roofs of boarding. Sails, windshaft and brake wheel were generally similar to those of the contemporary smock mill, but below the cap a completely novel arrangement of machinery was adopted. Instead of a wallower, the brake wheel drove a pinion mounted on a heavy iron crank shaft, which transmitted a reciprocating motion to a pair of multiple saw frames on the sawing floor below. Since the cap was fixed, it was necessary to turn the whole structure, including the logs to be sawn, to face the wind. The enormous weight was carried, in part by a massive central king post, and in part by rollers revolving on a low brick base. Ogee caps with circular portholes and boldly carved enrichments contributed to the bizarre appearance of these curious mills, whose name derived, like that of the British smock mill, from a fancied resemblance to traditional peasant costume.

Practical considerations limited the size of paltrok mills, and for heavier work the conventional smock mill was eventually adapted to sawing. This was done by interposing a short central shaft under the cap, driven from the brake wheel by a wallower, but geared to an iron crank shaft on the floor below. It was no longer necessary to restrict the size and weight of the structure, and a spacious timber sawing shed extending on either side of the tower characterises mills of this type. A hexagonal or square plan was usually adopted for the body of the mill, to ensure that the passage of timber through the saws was not impeded by cant posts.

Sawmills were normally built at the water's edge. In the paltrok mill, logs were floated alongside and lifted aboard by crane. In the smock mill, they were dragged up a long sloping ramp of boarding, and through double doors to the sawing floor. Here they were laid on sliding carriers, and propelled through the saw frame by rack and pinion gearing. The forward movement of the carrier was effected by a large ratchet wheel, turned a few degrees with each stroke of the saw by means of an ingenious system of levers and pawls.

These are a few of the basic machines employed over the centuries in harnessing the natural forces of wind and water. Inventive minds have found many other applications, and among the more curious was the

The Rat *at Ylst—this sawmill was transported to Friesland from the Zaan*

The Rat—*the crank shaft and connecting rod against the thatched walls of the mill*

FAR LEFT
The Rat—*the saw frame and carrier. When in use, the frames would hold a number of parallel saw blades—in the photograph one only remains in situ*

LEFT
The Rat—*the ratchet wheel and pawls governing the motion of the timber carrier: a saw blade is visible on the extreme right*

BELOW
The Rat—*an interior view, showing the timber carriers (foreground) and the saw frames. Winch barrels used to haul timber into the mill are visible at high level, behind the saw frames*

automatic spit. Meat roasting on spits before an open fire was traditionally turned by scullions, or by dogs trained to work a treadmill, poor 'turnspits' driven on by a hot coal placed in the bottom of the wheel. Weight-driven and clockwork spits became common in the eighteenth century, but many years earlier a very specialised form of windmill, the hot air turbine, had been developed for this work. A sketch by Leonardo da Vinci, dating from the late fifteenth century, shows a thoroughly practical design for such a machine, driven through gearing by a horizontal fan set in the throat of the flue above. A number of spits of this type, dating from the eighteenth and nineteenth centuries survive in the kitchens of great houses. The mechanism was simple, a worm mounted on the foot of the fan spindle driving a horizontal axle which projected through the chimney breast. On this was set a wooden pulley, which transmitted power through a belt to the spits below. The rising column of hot air from the fire provided a splendidly appropriate source of power; but even in the kitchen, the use of the water wheel was not unknown. At Mottisfont Abbey in Hampshire there remains a most unusual piece of machinery. The driving axle above the kitchen fireplace is linked not to a fan, but to a small breast-shot wheel, located in a cupboard and served by a leat running beneath the floor of the house.

A sawmill at Leiden

A hot air wheel for turning spits in Strangers' Hall, Norwich

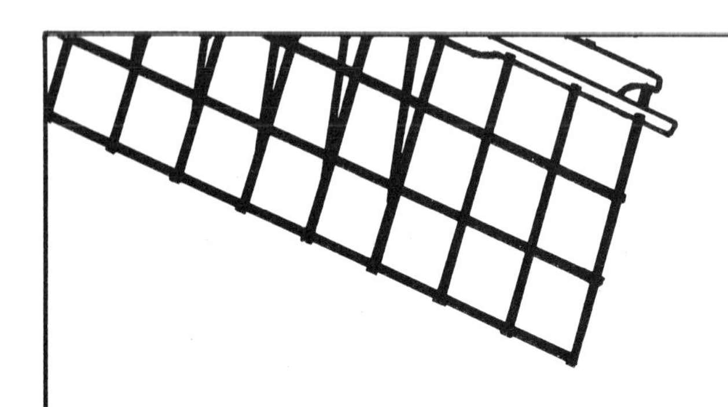

Conclusion

Since the Industrial Revolution, man has been using up the natural resources of the earth at an ever-increasing rate. Deposits of coal and oil, laid down through countless ages by the slow growth and decay of vegetation, dwindle yearly to feed the insatiable machines of the modern world, and the time is coming when man must look elsewhere for fuel. He has already paid a formidable price for the power generated during this brief period of accelerated industrial activity. The conversion of fuel to energy has been staggeringly wasteful, and the products of combustion have so polluted the atmosphere that his very existence on the planet is threatened.

By modern standards the traditional watermill and windmill were feeble and inefficient. But their use of natural power placed them in a class apart from the heat engine, and ethically they remain the most satisfactory machines devised by man. The unfailing reservoirs from which they drew their power were constantly renewed by forces beyond man's control or comprehension. To the medieval mind the mill at least approached the mystical ideal of perpetual motion, and would work on as long as the wind blew and the rain clouds passed overhead.

In fact, the principle of the watermill has never been discarded. Though the latest examples are not referred to as mills, and are devoted to a process unknown to the medieval millwright, they are nevertheless true lineal descendants of the primitive Greek wheel. The hydro-electric power station is a relatively modern development, which has resulted in some of the most impressive engineering works ever undertaken. These great schemes are costly, but the resulting benefits are incalculable. The power of the most awe-inspiring waterfalls has been harnessed to the service of man, and huge dams have been constructed to retain lakes exceeding a thousandfold the largest mill ponds of earlier centuries.

In the past, practical considerations have limited the general use of hydro-electric power to mountainous country, but the current renewal of interest in the principle of the tide mill presents a series of challenging possibilities. In a daring pilot project, French engineers have thrown a great barrage across the mouth of the river Rance, the first installation of its kind to produce electricity on a commercial scale. The ebb and flow of the tide offers an enormous potential source of power, and a number of similar projects are receiving active consideration on both sides of the Atlantic.

Scientists are now converting the energy of the sun to provide power for man's needs, and it is not perhaps too fanciful to classify the devices employed as solar mills. Like windmills and watermills, they follow the basic principle of the earliest engines, deriving their power from a natural source of energy, in this case the furnace of the sun.

Steam and internal combustion engines have been in existence for a brief period of time in comparison with the traditional watermill. They have proved to be indifferent servants, with unforeseen vices, and in the interests of his own survival, man must pay more attention to the development of natural power, conserving where he can the world's depleted stocks of fuel.

This book has outlined some of the ways in which wind and water power have been used in the past. The engines illustrated have served their turn and now stand idle. But modern turbines and Pelton wheels remain at work, producing power beyond the wildest dreams of the early millwright. They do not demand a daily tribute of coal or oil, and they do not poison the environment with harmful gases or radio-active waste. The water from their foaming spill-ways passes on to serve man's other needs. The hydro-electric power station may thus reasonably be regarded as the only truly civilised installation of its type. It is to be hoped that the future will see the increasing use of this admirable source of power, and that generations to come will spare a thought for the millwrights of past centuries whose work made these developments possible.

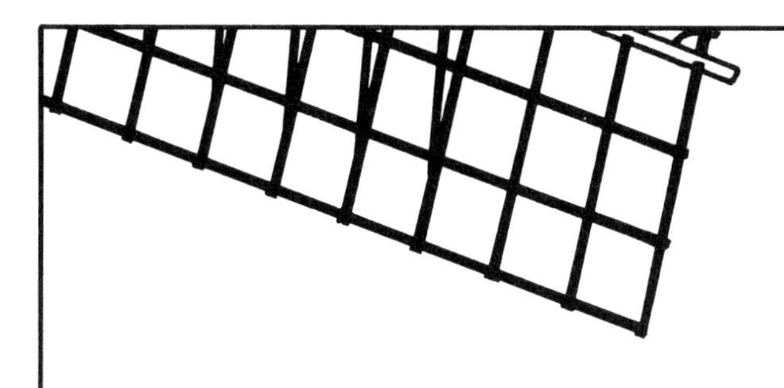

Glossary

Air brakes, longitudinal shutters sometimes incorporated in patent sails to reduce the speed of rotation in strong winds; also known as skyscrapers.

Air poles, conspicuous rod linkage forming part of the striking mechanism of Hooper's roller reefing sails.

Anchor chain, a chain used in conjunction with a windlass mounted on the tail pole of a windmill to turn (and secure) the mill; made fast to posts or bollards set around the base.

Annular sail, circular sail assembly with radial vanes.

Ark (miller's ark), a bin for the storage of meal.

Aruba penstock, a hollow tower of masonry designed to deliver a jet of water under pressure.

Auger, an Archimedean screw revolving in a wooden trough and employed for the horizontal movement of grain or meal; also known as conveyor or creeper.

Backstays, raking wooden braces providing support for the sail bars.

Back watering, the braking effect caused by the immersion of the lowest floats of an overshot water wheel.

Balance weights, weights inserted in the runner stone to achieve perfect balance. Patent balance weights employed discs of lead or iron carried on a threaded screw.

Beard, a decorative date board fitted under the weather beam in Dutch smock and tower mills.

Beatling, a pounding process in the manufacture of linen, performed by beatling stocks.

Bed stone, the lower, fixed stone in a pair of millstones; also known as ligger.

Bell alarm, a warning bell, triggered by lack of grain in the hopper; also known as warbler.

Bill, a hard steel cutting tool, in the form of a double ended wedge, used for dressing millstones; mounted in a turned wooden handle or thrift.

Bins, storage compartments for grain, usually arranged on the top floor of the mill (bin floor).

Bist, a small cushion of bran, used by the stone dresser.

Blowing house, a structure housing a furnace for smelting metallic ore and equipped with water powered bellows.

Blue stone, an imported German millstone.

Boat mill, see under floating mill.

Bobs, heavy counterweights, hung on bell cranks, and employed to balance the weight of the pump rod in mine drainage installations.

Bollard, the horizontal axle or barrel of a sack hoist.

Bolter, a device for dressing flour, employing a cylinder of cloth. This was at first of wool, latterly of silk, hence the term 'silks'.

Bouts, revolutions of wheels (or sails).

Brake, in windmills, a band brake of wood or iron working on the outer rim of the brake wheel, and controlled by a brake lever, or staff, and brake rope; also known as gripe.

Brake wheel, primary gear wheel in a windmill, consisting of a face wheel mounted on the windshaft; also known as head wheel.

Brayer (bray), pivoting beam or lever supporting the free end of the bridge tree.

Breast (of a millstone), the middle section of the grinding surface.

Breast (of a post mill), the fore part of the buck.

Breast beam, horizontal transverse beam supporting the neck of the windshaft; also known as weather beam.

Breast-shot wheel, a mill wheel driven by a positive head of water amounting to less than the diameter of the wheel.

Bridge tree, an adjustable beam supporting the thrust bearing at the foot of the stone spindle.

Bridging box, seating for a thrust bearing, incorporating hackle screws to permit lateral adjustment.

Brigging (brigging the spindle), the accurate plumbing of the spindle, by adjustment of the thrust bearing.

Bright, pasture glistening with water, but not flooded.

Buck, East Anglian term for the body of a post mill.

Buckets, floats, especially enclosed floats, employed in breast-shot and overshot wheels to retain water on the 'working' side of the wheel.

Burr stone, imported French millstone, of high quality, quarried in the Paris basin.

Cams, projecting studs on a trip wheel or axle (camshaft), employed to operate hammers or stamps.

Canister, an iron casting at the outer end of the windshaft to accommodate the sail stocks; also known as poll end.

Cant post, corner post of a smock mill.

Cap, the revolving top of a tower or smock mill.

Cap frame, horizontal timber frame forming the base of the cap.

Carding engine, combing device employed in woollen mills to produce a thick roll of wool prior to spinning.

Chain posts, bollards used for making fast the ends of the chains used in winding a mill equipped with a tail pole; also known as anchor posts.

Clamps, wooden splints bolted to the sides of the stocks, locking the sail assembly in the poll end.

Clapper, a block of wood or stone used as a primitive regulator to control the flow of grain to the stones.

Clasp arm wheel, a timber wheel built up on a cruciform frame consisting of two pairs of parallel spokes enclosing the axle.

Cock head, the upper termination of a spindle or shaft.

Cog hole (cog pit), in watermills, the space enclosed by the hursting, containing the main gearing.

Common sail, the traditional windmill sail of northern Europe, covered with a canvas 'cloth'.

Compass arm wheel, a timber wheel with radiating spokes, mortised to the axle.

Composite mill, a rare variant of the post mill, in which the weight of the body is carried on the walls of the roundhouse.

Composition stone, artificial millstone containing carborundum or similar abrasive material.

Conveyor, see under auger.

Cow-pop gear, a face wheel fitted with projecting wooden pegs; also known as trundle wheel.

Cracking, a process in stone dressing involving the cutting of fine grooves in the surface of the lands; also known as stitching.

Creeper, see under auger.

Crook string, a line controlling the angle of the shoe feeding grain to the stones.

Cross, cruciform iron support for the sail assembly, used as an alternative to the poll end.

Cross-trees, main horizontal members of the underframe of a post mill.

Crown tree, principal framing member of the body of a post mill; a stout transverse beam, pivotting on top of the main post.

Crown wheel, in watermills, a bevel gear wheel mounted near the top of the main vertical shaft.

Cullin stones (Cologne stones), imported German millstones.

Curb, in smock and tower mills, the circular timber rim or wall plate, supporting the revolving cap.

Dagger point, a degree of reefing employed with common sails: refers to the pointed outline assumed by the outer end of the sail cloth.

Damsel, an iron rod terminating in a fork or crutch which straddled the rynd of the runner stone: employed with under-driven machinery to agitate the shoe feeding grain to the stones; also known as dandelion.

Dead curb, a curb supporting the cap without the interposition of wheels or rollers.

Dekkerise (after A. J. Dekker, a Dutch millwright), the modification of windmill sails by the addition of aerodynamically designed metal fairings enclosing the leading edge and stock.

Descender, a device invented by Oliver Evans for the controlled downward movement of grain within the mill.

Double shuttered, patent sails or spring sails fitted with shutters on both sides of the whip.

Drag stick, a stick inserted in the eye of the runner stone, to prevent the hoarding (or clogging) of grain.

Drainage mill, a windmill equipped for land drainage.

Dresser, a device for grading flour, consisting of a cylindrical wire sieve containing rotating brushes; also known as wire machine.

Dressing (of stones), the preparation of grinding surfaces.

Driving stocks, water powered tilt hammers employed for fulling cloth.

Dust floor, the top floor in a tower or smock mill.

Dust rooms, sealed compartments for the collection of dust extracted during the milling process.

Dutch blue, see under blue stone.

Edge runner stones, stones designed to rotate in a vertical plane, the edge forming the crushing surface: employed in various industrial processes.

Edge tool works, forges specialising in the production of cutting implements, such as scythes.

Elevator, a device for raising grain by means of an endless belt, invented by Oliver Evans.

Ending stones, a pair of stones set well apart, employed in American mills to crack the bran as a preliminary to 'low' grinding.

Ends, the method used in Holland to record the speed of a windmill by counting the sails (or ends); thus one revolution of the sail assembly equals four ends.

Eye, the central hole through the runner stone.

Face wheel, a gear wheel cogged on the face, used in conjunction with a pinion.

Falling stocks, water powered stamps for fulling cloth, driven by a camshaft.

False board, upper section of the inclined hatch or sluice gate, used with undershot wheels of Poncelet type.

Fan stage, the platform giving access to the fantail from the back of the cap, in tower or smock mills.

Fang staff, brake lever.

Fantail, the wind wheel set at right angles to the sails, which provides the motive power for automatic winding gear; also known as fly.

Felloe (felly), section of the rim forming part of a built-up wooden wheel.

Fer-de-moline, heraldic term for the mill rynd.

First reef, the first degree of reefing with common sails; see also dagger point.

Flash lock, a primitive forerunner of the pound lock: adjustment of a removable stanch or sluice released a flash of water, on which boats could be hauled upstream.

Flat rods, iron rods employed for the horizontal transmission of power in cases where machinery was unavoidably sited at a distance from the water wheel.

Floating mill, a vessel containing milling machinery, moored in midstream and powered by one (or two) undershot wheels; also known as boat mill.

Floats, blades of a water wheel.

Flood hatch, a hatch diverting water through the spillway, clear of the water wheel.

Flume, head race in the form of a trough or channel carrying water to the wheel; also known as lade.

Footbrass, the thrust bearing for an upright shaft or spindle; also known as footstep bearing.

Frames, in stamp mills, the timber structure enclosing the stamps.

French stones, see under burr stone.

Full sail, common sail with cloth fully spread.

Fulling stocks, water powered mallets or stamps employed for fulling cloth.

Furrows, grooves cut in the grinding surfaces of millstones.

Gallery, stage around the cap of a smock or tower mill.

Gimbal bar, see under rynd.

Glut box, top bearing of the quant,which could be opened to disengage the stone nut.

Governor, a regulator, usually of the centrifugal type, employed in windmills for the automatic adjustment of the clearance between stones to suit the speed of the wind.

Gradual reduction, the production of flour by multiple grinding.

Great spur wheel, spur wheel mounted on the main vertical shaft, transmitting the drive to two or more pairs of stones.

Grist, material ground in the mill: latterly applied principally to animal feeding stuffs.

Groats, roasted and husked oats: husks removed in the groat machine.

Grutte, a bearing inserted in the bed stone of the Norse mill to receive the upper end of the vertical spindle.

Gudgeon, a metal journal mounted in the end of a wooden shaft or axle: cross head gudgeons were frequently used in mill work, in conjunction with iron hoops.

Hackle plate, the square cover plate screwed down over the spindle bearing in a bed stone.

Hammer pond, mill pond associated with a forge, supplying power for tilt hammers.

Harp, segment of the grinding surface of a millstone, containing a regular pattern of lands and furrows.

Head and tail, common arrangement of stones in the post mill, permitting two pairs to be driven without the interposition of a vertical shaft.

Head race, that section of the mill stream above the wheel.

Head sick, an expression describing a post mill out of balance, with the body tilting forwards.

Helve, the shaft of a tilt hammer.

Hemlaths, light longitudinal members forming the edges of the sail frame.

High grinding, the process of milling by gradual reduction.

Hollander (Hollander beating engine), a device employed for pulping rags by means of a revolving drum equipped with blades. Superseded stampers in paper mills.

Hollow post mill, a windmill in which power was transmitted through a hollow central post to machinery housed in the base; known in Holland as *wip* mill.

Hoodway, wooden casing enclosing the scoop wheel of a drainage mill.

Hopper, an open wooden container for the controlled discharge of grain to the millstones.

Hopper boy, a device invented by Oliver Evans for spreading freshly ground meal to cool prior to bolting.

Horizontal air mill, a windmill powered by vanes or sails mounted on a vertical shaft.

Horizontal mill, a watermill driven by a horizontal rotor mounted on a vertical shaft; also known as Greek or Norse mill.

Horse, wooden framework supporting the hopper above the millstones.

Hull (grinding hull), a small mill specialising in the grinding of blades.

Hulling mill, a mill equipped for the husking of cereal grains such as barley and rice.

Hunting cog, a cog deliberately introduced to avoid simple gear ratios, thus ensuring even wear.

Hursting, timber framework supporting the millstones, and enclosing the main gearing in the watermill.

Inside winder, a tower or smock mill with winding gear consisting of a windlass mounted inside the cap; known in Holland as *binnenkruier*.

Irrigation wheel, see under noria.

Jib sails, triangular windmill sails of cloth, set on radiating spars.

Jog-scry, an oscillating sieve.

King post, main central post of a paltrok sawmill.

Lade, see under launder.

Lands, surface of a millstone between the furrows.

Lantern pinion, an early form of pinion consisting of staves mortised between wooden discs.

Launder, a trough carrying water to the wheel: also known as lade.

Leading board, the board fitted along the leading edge of the sail, directing the flow of air on to the cloth (or shutters).

Leat, mill stream.

Lifters, the wooden shafts of a stamp mill.

Ligger, see under bed stone.

Lightening tree, an adjustable wooden rod supporting the end of the sole tree in the Norse mill.

Lighter screw, an adjustable screw connecting the brayer to the steelyard: part of the tentering linkage in the windmill.

Live curb, the curb supporting a windmill cap revolving on rollers or wheels.

Locum, oversailing dormer enclosing the hoist in a watermill.

Logwood mill, a mill equipped for rasping logwood, used to produce a dark red dye.

Loper, see under runner stone.

Low milling, the traditional method of grinding, reducing grain to flour in a single passage through the stones.

Luff, to bring the mill head to wind.

Mace head, the head mounted on the upper end of a stone spindle, supporting (and driving) the gimbal bar (or rynd) of the runner stone.

Magnetic separator, a device introduced in the latter part of the nineteenth century to remove stray metal (binding wire, etc.) from grain prior to grinding.

Makelaar, decorative finial carrying a wind vane, fixed at the gable end of Dutch windmills.

Meadow mill, a small, untended drainage mill.

Meal floor, the floor below the millstones, to which the freshly ground meal was discharged.

Meal house, the upper chamber of the Norse mill, containing stones and hopper.

Meal spouts, wooden spouts conveying meal from the stones.

Middling, see under stock.

Miller's willow, a wooden spring employed to tension the shoe against the damsel or quant.

Mill-in-a-mound, a tower mill with a mound built up around the base to serve in place of a stage; known in Holland as *bergmolen* or *beltmolen.*

Mood, a billet of laminated steel for forging into blades.

Multure bowl, the receptacle used by the miller for taking toll; also known as toll dish.

Neck, front journal of the windshaft, revolving in the neck bearing.

Noria, a vertical water wheel, turned by the current, lifting water for irrigation by means of containers attached to the rim; also known as Persian wheel and wheel of pots.

Norse mill, the term employed in Northern Europe for the Greek or horizontal mill.

Norway engine, an early term for the water powered frame saw.

Nose helve, a tilt hammer lifted by the head, in the manner of fulling stocks.

Oil mill, a mill equipped with edge runner stones and stamps, for the extraction of vegetable oils.

Oliver, a treadle operated hammer, with sapling return spring.

Onker, a sailing vessel, usually a barque, employed during the nineteenth century in the Scandinavian timber trade.

Outside winder, a tower or smock mill with the cap turned by a tail pole; known in Holland as *buitenkruier.*

Overdrift, millstones driven from above, by quants.

Overshot, water wheel served by a head race discharging over its outer circumference.

Paint staff, a wooden straight edge used with raddle to test the surface of a stone.

Pair, a term used to describe the capacity of a mill by reference to the number of pairs of millstones installed; for example, 'two pair mill', 'three pair mill' and so on: see also run.

Paltrok mill, the earliest type of wind driven sawmill, introduced in Holland at the end of the sixteenth century.

Pandy, Welsh term for a fulling mill.

Patent sail, a sail fitted with shutters controlled by automatic striking gear.

Peak stone, British millstone of 'millstone grit' from Derbyshire.

Peg mill, a post mill with buried underframe; also known as sunk post mill.

Penstock, a wooden trough carrying water to the wheel, or alternatively a sluice gate controlling the flow.

Petticoat, the cladding of a post mill carried down to weather the central opening in the roof of the roundhouse.

Pick, a pointed bill used for stone dressing.

Pinch bar, a device related to the mason's lewis, used to lift the runner stone for dressing or renewal.

Pintle, the journal forming the upper termination of a post or spindle.

Pit wheel, primary gear wheel of the watermill, mounted on the inner end of the axle.

Pitch back wheel, a water wheel of the overshot type, with the launder modified to discharge on to the near side of the wheel, thus reversing the normal direction of rotation.

Plansifter, a machine equipped with horizontal sieves for dressing flour.

Plating hammer, a tilt hammer used for drawing out blades.

Plim, to swell cogs or wedges with water in order to counter the loosening effect of vibration.

Pointing lines, control lines attached to the edges of sail cloths, allowing the miller to shorten sail from the ground below.

Polder mill, Dutch drainage mill of the smock type.

Poll end, see under canister.

Poncelet wheel, an improved type of undershot water wheel, fitted with curved metal floats, invented by General J. V. Poncelet.

Post mill, the earliest form of European windmill, the body revolving on a stout upright post.

Prick post, central stud in the breast of the post mill.

Proof staff, iron straight edge, used to check the paint staff.

Purifier, a device introduced early in the nineteenth century to extract fine dust from the meal.

Quant, the driving spindle used with overdrift stones.

Quarterbars, members of the post mill underframe: raking struts supporting the main post.

Quartering, the action of turning the mill broadside to the wind.

Rack, bar or rod cogged to engage with a pinion or worm; specifically the cogged track fixed around the curb in mills equipped with mechanical winding gear.

Raddle, paint composed of red oxide and water, applied to the proof staff to detect and mark raised areas on grinding surfaces of stones prior to dressing; also known as tiver.

Rag and chain pump, a lift pump employing an endless chain carrying stuffed balls of leather or cloth.

Rams, stamps consisting of vertical baulks of timber, used to drive and release pressing wedges in oil mills and similar industrial installations.

Rap, fender protecting the wooden side of the shoe at the point where it makes contact with the damsel.

Recoil block, a resilient block mounted under the tail of a tilt hammer shaft to limit the length of the stroke; sometimes bedded on compressed heather.

Reel, the silk covered cylinder of the bolter.

Riggers, pivotting iron levers, with forked ends, used to lift iron stone nuts out of gear.

Rind or rynd, an iron bearer set across the eye of the runner stone; more correctly described, in later mills, as gimbal bar.

Roller mill, a mill employing metal rolls in place of the traditional stones.

Roller reefing sails, sails fitted with canvas strips wound on rollers in place of shutters.

Roundhouse, circular structure enclosing the underframe of a post mill.

Run (of stones), American term for a pair of stones, used to describe the capacity of a mill; see also pair.

Rungs, the floats of a water wheel; also, in windmills, the transverse iron rods carrying the sail cloths.

Runner stone, the upper, revolving millstone.

Sack boy, a device for holding open the mouth of a sack.

Saddle quern, a primitive hand mill employing reciprocating motion.

Sail back, a single spar replacing the stock and whip, in mills fitted with iron crosses in place of poll ends.

Sail bars, transverse members of the sail frame.

Sampson head, an iron pivot at the head of the main post, carrying the body of a post mill.

Scoop wheel, a narrow paddle wheel with raking floats, used for lifting water in drainage mills.

Screener, a machine for removing dust from grain prior to grinding.

Scribbling, a preliminary carding or combing process employed in the preparation of wool.

Separator, a machine used in place of, or in conjunction with, a screener.

Sheers, in post mills, the principal framing members of the floor of the buck running fore and aft on either side of the post: in smock or tower mills, the principal longitudinal members of the cap frame.

Shoe, a tapering wooden trough, fed with grain from the hopper and discharging to the eye of the runner stone.

Shot curb, a form of live curb supporting a ball race under the revolving cap of a smock or tower mill.

Shroud, a deep rim enclosing the buckets in overshot or breast-shot water wheels.

Shutter bar, a longitudinal rod operating a group of shutters.

Shutters, pivoting vanes, used in spring and patent sails, in place of the canvas 'cloths' of the traditional common sail.

Sickle dress, a method of stone dressing employing curved radiating furrows.

Side girts, horizontal framing members to the sides of the post mill body.

Sile, rynd as used in the Norse mill.

Silks, see under bolter.

Single shuttered, a sail having shutters on the trailing or driving side of the whip only.

Skaltkvarn, Scandinavian term for the horizontal or Norse mill.

Skirt, the outer section of the grinding surface of a millstone.

Skirting board, the low wooden curb scribed around the bed stone at floor level.

Skyscrapers, see under air brakes.

Slip cogs, removable cogs, secured by iron pins, permitting a pinion to be taken out of gear.

Slubbing billy, a water powered spinning engine, developed from the Spinning Jenny.

Smock mill, timber framed counterpart of the tower mill; clad in weather boarding, and named after a fancied resemblance to the countryman's white linen smock.

Smutter, a machine for removing smut, a fungal growth common in wheat.

Soke (milling soke), ancient custom regulating the conduct of milling.

Sold, a miller's sieve of punctured whaleskin, used in the Faroes.

Sole, timber or sheet iron lining to a water wheel; a board forming the bottom of an individual bucket.

Sole tree, a pivoting beam carrying the thrust bearing of the vertical spindle in the Norse mill.

Spattle, a sliding shutter controlling the flow of grain from hopper to shoe.

Spider, the cast iron, star-shaped frame for a wheel, comprising hub and radiating spokes; also the cruciform linkage at the centre of the patent sail assembly, connecting the striking rod to the shutter bars.

Spider mill, diminutive hollow post mill, used for drainage work in Holland.

Sprattle beam, horizontal beam carrying the sprattle box, or top bearing of the vertical shaft in tower or smock mills; also known as centre beam.

Spring sail, a windmill sail fitted with shutters controlled by springs.

Staff, see under proof staff.

Stage, the projecting gallery around tall tower or smock mills, providing access to the sails and tail pole (where retained); also known as reefing stage.

Stampers, stocks used in the process of paper making.

Stamps, iron shod baulks of timber employed as pestles in various industrial mills.

Star wheel, iron centre carrying the spokes of the fantail.

Starts, short spurs of wood or metal, projecting from the rim of a water wheel to support the floats.

Steady bearing, the bearing formed between the sheers at floor level in a post mill, to prevent the body of the mill from swaying; also known as girdle.

Steeling hammer, a tilt hammer used for welding billets of laminated steel.

Steelyard, an iron lever linking the governor to the brayer in a windmill.

Stitching, see under cracking.

Stock, a tapered spar passing through the poll end, and supporting a pair of sails.

Stone nut, pinion transmitting the drive from the spur wheel to the runner stone.

Storm hatch, an access hatch, located above the neck bearing of the windshaft, normally closed by a sliding shutter.

Strike (strickle), the straight edge used to level off grain or meal in a toll dish or measure.

Striking gear, the mechanism used with patent sails to apply pressure to the shutters, comprising a striking rod passing through the length of the windshaft, operated by an endless chain on which weights were hung to suit the force of wind; striking gear was also employed with roller reefing sails.

Sunk post mill, see under peg mill.

Suspension wheel, a lightly built iron framed water wheel, designed to drive through a cogged track fixed to the rim.

Swallow, American term for the eye of the runner stone.

Sweep, a Kentish term for the windmill sail.

Swing pot, a pivoting neck bearing sometimes used to support the windshaft.

Sword, the key used in the Norse mill to secure the upper end of the lightening tree.

Sword point, a degree of reefing employed with common sails; see also dagger point.

Tail, the rear of the post mill body (as opposed to the breast).

Tail helve, a tilt hammer driven by cams acting on the tail of the shaft.

Tail pole, the projecting spar used to turn a windmill not equipped with automatic winding gear.

Tail race, that section of the mill race downstream of the wheel.

Tail wheel, a gear wheel mounted near the tail of the windshaft in post mills, to drive a second pair of stones.

Tail-winded, a windmill caught aback by a sudden change of wind, exerting pressure on the wrong side of the sail assembly.

Talthur (tiller), the lever mounted on the tail pole of a post mill, to raise the ladder clear of the ground prior to winding the mill.

Temse, a sieve for dressing flour.

Tentering frame, a wooden rack on which cloth was spread after fulling.

Tentering gear, the mechanism for making fine adjustments to the gap between millstones.

Threshing mill, a farm mill equipped to drive a stationary threshing machine.

Thrift, the turned wooden handle for a bill.

Tide mill, a watermill utilising the rise and fall of the tide, and normally provided with a mill pond in the form of a tidal pound or reservoir.

Tilt hammer, an automatic hammer operated by a trip wheel or camshaft.

Tirl, the impeller or rotor of the Norse mill.

Tiver, see under raddle.

Tjasker, a simple Dutch drainage mill, consisting of an Archimedean screw driven by sails, without gearing.

Toll, payment in kind, taken by the miller in a toll dish or multure bowl.

Tower mill, a windmill of masonry or brickwork, fitted with a revolving cap.

Trammel, a wooden gauge used to check the plumbing of the stone spindle.

Tramp iron, American term for metal scrap (binder wire, etc.) hidden in the grain.

Truck wheels (truckles), small wheels revolving in a horizontal plane against the inner face of the curb, to centre the cap of a tower or smock mill.

Trundle wheel, a primitive gear wheel with projecting pegs as teeth; related to the lantern pinion.

Tub wheel, horizontal mill wheel revolving in a circular wooden casing.

Tucking mill, another term for fulling mill.

Tun, removable wooden casing enclosing the millstones; also known as vat or hoop.

Turbary, the practice of cutting peat for fuel; in legal terms an easement to dig turf or peat on another's land.

Twist peg, adjustable winding peg controlling the crook string.

Underdrift, millstones driven from below; the customary arrangement in the watermill.

Under-house, the lower compartment of the Norse mill, containing the tirl.

Uplongs, longitudinal battens in the sail frame assembly.

Vanes, see under shutters.

Vat, see under tun.

Ventilating buckets, water wheel buckets designed with vents to ease the entry and release of water.

Vertical mill, the traditional European mill driven by a vertical water wheel.

Vitruvian mill, the Roman vertical mill, as described by Vitruvius.

Wall mill, a windmill set on the ramparts of a castle or fortified town.

Wallower (waller), the bevel gear wheel or lantern pinion, driving the main vertical shaft in both watermill and windmill.

Weather beam, see under breast beam.

Weather (of sail), the angle or pitch of the sail.

Whip, principal longitudinal framing member of the individual windmill sail, strapped and bolted to the face of the stock.

Winding, the action of turning the mill head to wind.

Windshaft, the axle carrying the sail assembly.

Winter mill, a class of mill mentioned in the Domesday survey, probably powered by small seasonal streams.

Wip mill, Dutch term for the hollow post mill.

Wire machine, see under dresser.

Worm, cylindrical gear bearing a helical thread; frequently used in conjunction with a rack in windmill winding gear.

'Y' wheel, a wooden pulley with iron crutches projecting from the rim, for operation by an endless rope or chain.

Yoke, wooden bars projecting downwards from the tail pole, against which the miller set his shoulders when winding the mill.

Yoke (of edge runner stones), the timber mounting for a pair of stones, arranged to revolve around the foot of the vertical drive shaft.

Zaan, an industrial district of North Holland, notable for the variety and number of its mills; Zaan Green, a characteristic colour used for paintwork.

Zoetrope, a wheel or fan operated by a rising current of hot air.

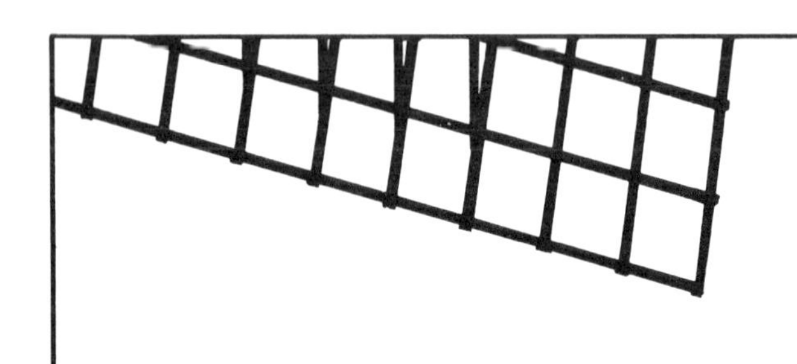

Selected Bibliography

In gathering information for this book the author has drawn upon a number of sources. The following is a personal and selective reading list compiled by the author.

Arnold, J. The Shell Book of Country Crafts: London, 1968

Ashton, T. S. Iron and Steel in the Industrial Revolution: London, 1924

Bennett, R. & Elton, J. History of Corn Milling (4 volumes): London, 1898-1904

Booker, F. Industrial Archaeology of the Tamar Valley: Newton Abbot, 1967

Brangwyn, F. & Preston, H. Windmills: London, 1923

Brett, C. A Byzantine Watermill: *Antiquity*: Gloucester, 1939

Burstall, A. F. A History of Mechanical Engineering: London, 1963

Chaloner, W. H. & Musson, A. E. Industry and Technology, A Visual History: London, 1963

Crankshaw, W. P. Report on a Survey of the Welsh Textile Industry: Cardiff, 1927

Crossley, F. H. Timber Building in England: London, 1951

Cruden, S. Click Mill, Dounby, Orkney: Edinburgh, 1949

Curwen, E. C. The Problems of Early Watermills: *Antiquity*: Gloucester, 1944
A Vertical Watermill near Salonika: *Antiquity*: Gloucester, 1945

Derry, T. K. & Williams, T. I. A Short History of Technology from the Earliest Times to A.D. 1900: Oxford, 1960

Evans, G. E. The Farm and the Village: London, 1969

Finch, W. Coles Windmills and Watermills: London, 1933

Freese, S. W. Windmills and Millwrighting: Cambridge, 1957

Gardner, E. M. Tide Mills, Part 3: London, 1957

Hills, R. L. Machines, Mills and Uncountable Costly Necessities: Norwich, 1967

Hopkins, R. T. Old English Mills and Inns: London, 1927

Hopkins, R. T. & Freese, S. In Search of English Windmills: London, 1931

Hsing, Sung Ying Tien Kung K'ai-Wu, Chinese Technology in the Seventeenth Century (translated by E-Tu Zen Sun & Shiou-Chuan Sun): Pennsylvania, 1966

Hudson, K. Industrial Archaeology: London, 1966

Husa, V., Petran, J. & Subrtova, A. Traditional Crafts and Skills: London, 1967

Jenkins, J. Geraint The Esgair Moel Woollen Mill: Cardiff, 1965

Lilley, S. Man, Machines and History: London, 1948

Lipson, E. History of the Woollen and Worsted Industries: London, 1965

Long, G. The Mills of Man: London, 1931

Major, J. Kenneth Finch Brothers Foundry, Sticklepath: Newton Abbott, 1967

Pelham, R. A. Fulling Mills. A Study in the Application of Water Power to the Woollen Industry: London, 1958
The Old Mills of Southampton: Southampton, 1963

Percival, A. Faversham's Gunpowder Industry: Faversham, 1967

Reid, K. C. Watermills and the Landscape: London, 1959

Rhodes, P. S. Ballycopeland Windmill, County Down: Belfast, 1962

Shorter, A. H. Water Paper Mills in England: London, 1966

Singer, C., Holymard, E. J., Hall, A. R. & Williams, T. I. (Editors) A History of Technology: Oxford, 1956

Stockhuyzen, F. The Dutch Windmill: London, 1962

Storck, J. & Teague, W. D. Flour for Man's Bread: Minneapolis, 1952

Syson, L. British Water-mills: London, 1965

Vince, J. N. T. Discovering Windmills: Tring, 1965

Wailes, R. The English Windmill: London, 1954
Berney Arms Mill: London, 1957
Saxtead Green Mill, Suffolk: London, 1960
Tide Mills, Parts 1 and 2: London, 1956 & 1957

White, L., Jr. Medieval Technology and Social Change: Oxford, 1962

Williamson, K. Horizontal Water Mills of the Faeroe Islands: *Antiquity:* Gloucester, 1946
The earliest Engine: *The Countryman:* Burford, 1963

Wilson, P. N. Water-Mills with Horizontal Wheels: London, 1960

Winton, W. Iron and Steel: Science Museum, London

The following pamphlets deal with particular mills or aspects of milling. Since they do not bear the names of their authors, they are listed under the name of the mill or locality described.

Abbeydale Industrial Hamlet Council for the Conservation of Sheffield Antiquities

Bembridge Windmill National Trust, 1962

Calbourne, Isle of Wight The Water Mill

Dutch Windmills. Twelve Tours Dutch Windmill Society, the Hague

Horsey Drainage Mill National Trust, 1966

Kennet and Avon Canal, Pumping Stations Kennet and Avon Canal Trust Ltd

Kinderdyke Mills Mun. Nieuw Lekkerland

Oil Milling Unilever Ltd., 1955

Shepherd Wheel, Sheffield Council for the Preservation of Sheffield Antiquities

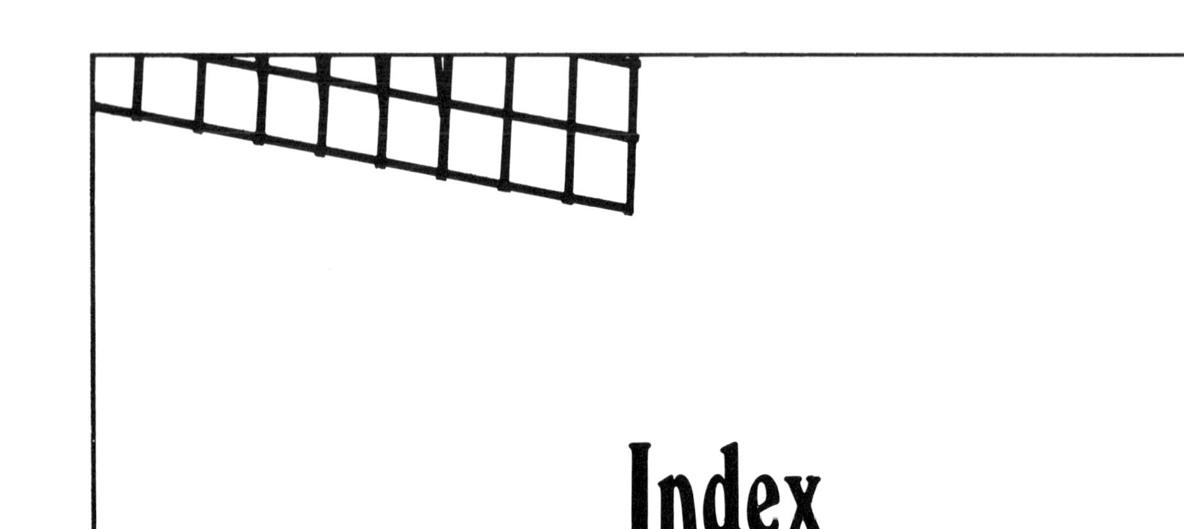

Index